EX 3

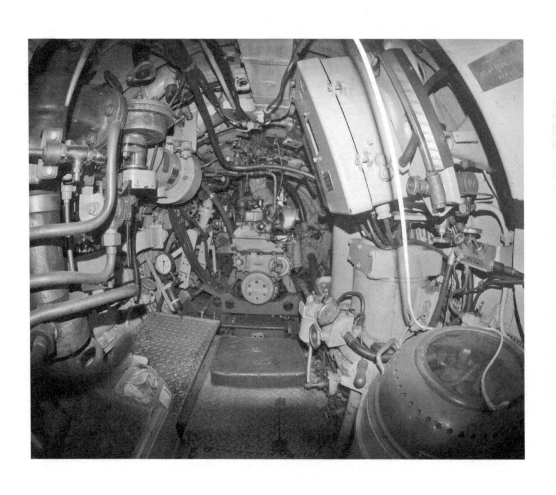

X3 TO X54
THE HISTORY OF THE BRITISH MIDGET SUBMARINE

KEITH HALL

All royalties from the sale of this book are being donated to the Submarine Centre, Helensburgh.

Front cover image: XE8 under way.
(*National Museum of the Royal Navy*)

Back cover image: Remembrance Day poppies.
(*Submarine Centre, Helensburgh*)

Frontispiece: The engine room, looking aft.
(*Submarine Centre, Helensburgh*)

First published 2023

The History Press
97 St George's Place, Cheltenham,
Gloucestershire, GL50 3QB
www.thehistorypress.co.uk

© Keith Hall, 2023

The right of Keith Hall to be identified as the Author of this work has been asserted in accordance with the Copyright, Designs and Patents Act 1988.

All rights reserved. No part of this book may be reprinted or reproduced or utilised in any form or by any electronic, mechanical or other means, now known or hereafter invented, including photocopying and recording, or in any information storage or retrieval system, without the permission in writing from the Publishers.

British Library Cataloguing in Publication Data.
A catalogue record for this book is available from the British Library.

ISBN 978 1 80399 199 3

Typesetting and origination by The History Press
Printed and bound by TJ Books Limited, Padstow, Cornwall

CONTENTS

Preamble — 7

PART ONE: X CRAFT HISTORY

1. The Mediterranean — 13
2. X Craft Development — 19
3. 12th Flotilla Operations — 29
4. X Craft Built During the Second World War — 49

PART TWO: THE ORIGIN OF THE STICKLEBACK CLASS

5. The Cold War — 57
6. Britain after the Second World War — 69
7. The Stickleback Class Introduction — 73

PART THREE: BUILDING THE STICKLEBACK CLASS

8. Constructing the Stickleback Class — 81
9. Layout — 89
10. How the X Craft Works — 111
11. Life On Board the Submarine — 113
12. Stickleback Class Operational Life — 117

PART FOUR: THE SUBMARINE CENTRE, HELENSBURGH

13. The Submarine Centre, Helensburgh — 137

Epilogue — 163
Bibliography — 171
Acknowledgements — 173
Also by the Author — 175

PREAMBLE

The Royal Navy developed several midget submarines during the Second World War. However, the decommissioning of the Stickleback class between late 1958 and early 1960s marked the end of the Navy's enthusiasm for this type of craft. They were initially developed for one purpose and one purpose only, to attack major enemy ships in harbour; however, their usefulness and adaptability was quickly recognised and the range of their missions grew. These craft formed the 12th Flotilla during the war, which included X craft and manned torpedoes. These craft were based at Port Bannatyne on the Isle of Bute.

Submersible vehicles may have been used for military purposes as early as the fourth century BCE, when, legend has it, Alexander the Great used a glass diving bell for reconnaissance. The Middle Ages gave birth to many designs for exploratory and military submersibles, most of which, luckily for the potential crews, were never built.

During the America Revolutionary War in 1775, Connecticut-born David Bushnell, who was a student at Yale in the early 1770s, studied the use of underwater explosives and how they could be used to attack British ships. He developed this idea and, with funding provided by General George Washington, an egg-shaped submersible was produced. This became known as the Turtle; it was a hand-powered craft that used a hand-powered drill and a ship's auger to attach explosives to the hull of an enemy ship.

This very small craft was 10ft by 3ft and had a crew of one man, who propelled his submarine by pedalling. It was made of wood and covered with tar with steel bands for reinforcement, similar to a barrel. On the surface, several small windows allowed the crewman to navigate. A small water tank could be flooded to enable the Turtle to submerge; to surface the submarine, the water could be pumped out by hand.

On 7 September 1776, this unique craft was used to attack General William Howe's flagship, HMS *Eagle*, in New York Harbor. General

Washington gave permission for this mission and at 11 p.m. on 6 September, Sergeant Ezra Lee pedalled for two hours in his approach to *Eagle*. Although the attack was unsuccessful, this was the first naval attack ever made in a submarine.

The Turtle made several attempts to attach explosives to the hulls of the British warships anchored in New York Harbor during 1776, but none were successful. A few months later, the British sunk the submarine's 'depot ship' while it was on board. Although Bushnell claimed to have salvaged his submarine, the claim was never confirmed. Washington later wrote that the invention was ingenious, but contained too many variables to be controlled, a thought echoed by many submariners even to this day. A replica of the Turtle is on display in the Royal Navy Submarine Museum at Gosport.

As the Americans realised, enemy ships were more vulnerable to attack in harbour than when they are at sea. During the Second World War, enemy capital ships were anchored in their northern, heavily defended lairs. For aircraft, there were limited and well-defended approaches; anti-aircraft guns and smoke generators made the targets almost invulnerable. Despite this, naval authorities thought these vessels remained vulnerable to underwater attack if the right submersible was available.

The Second World War British X craft were developed to exploit this weakness; the post-war X craft were constructed to carry out a different task. The latter were built to test the country's harbour defences, to see if they could meet a challenge posed by the perceived Russian midget submarines.

Part One of the book explores the history of the X craft, describing their development and the Second World War operation they were involved in. Part Two details the origins of the Stickleback class. Part Three covers the building of this new class of midget submarine and its layout, examines the rather tough life on board one of these very cramped and uncomfortable craft, and looks at their rather short operational life.

Part Four covers the Submarine Centre in Helensburgh and the bringing of HMS *Stickleback* to Scotland. Apart from highlighting the town's close links with the submarine community, one of the founder's main aims was to produce an exhibition that would act as a memorial to the men of the submarine service and the crews of the 12th Flotilla in particular. The centre's hi-tech presentation certainly achieves this.

The Epilogue covers other memorials to the X craft and where the few remaining ones can be seen.

PREAMBLE

I have purposely avoided including a glossary and I explain any abbreviations in the text, hopefully avoiding the need for the reader to flip backwards and forwards through the book. For a similar reason, I have not included footnotes or references.

I hope you enjoy the book.

Keith Hall

PART ONE

X CRAFT HISTORY

1

THE MEDITERRANEAN

In the early part of the Second World War, the Italians realised they couldn't match the Royal Navy, and Mussolini's efforts to seize control of the Mediterranean had, by late 1941, failed. The raid on Taranto and the Battle of Cape Matapan had given the Royal Navy a decisive advantage. Low morale and fuel shortages further impeded the Regia Marina's capabilities. Nevertheless, the Italians still had several modern battleships and a few older, modernised vessels. Fortunately for the Italians, the Royal Navy had lost one of its battleships; a German U-boat had sunk HMS *Barham* on 25 November 1941 and the British had deployed several warships to the Far East in expectation of war with Japan.

Just before the outbreak of war, the Italians set up a special unit to develop innovative ideas to level the balance of power with the Royal Navy. In 1941, this unit became the 10th Flotilla (Xa MAS). They developed two primary/novel methods of attacking the British Mediterranean Fleet.

The first was a 2-ton, 18ft-long wooden speedboat packed with 728lb of explosives. It was powered by an Alfa Romeo AR 6cc Outboard Motor rated at 95hp and could reach speeds of 33 knots. It was known as the Motoscafo da Turismo (MT) and nicknamed *Barchino* (little boat). These boats were specially designed to make their way through obstacles such as torpedo nets. Once the operator aimed the boat at his target, he locked the steering and dived into the water.

The second weapon, which resulted from a collaboration between two young engineers in the Italian Navy's Submarine Service, was the human torpedo, known as *siluri a lenta corsa* (SLC, or slow-speed torpedo).

This was not an entirely new idea. In 1918, days before the surrender of the Central Powers, the collapsing government of Austria-Hungary made clear that it would pass one of its dreadnought battleships, SMS *Viribus Unitis*, on to the newly created State of Slovenes, Croats and Serbs, a forerunner of Yugoslavia. The Italians did not want to see a powerful battleship falling into the possession of a potential regional rival; Italy believed the ship still belonged to Austria-Hungary and no peace treaty yet existed. Accordingly, a team of divers entered Pula Harbor and attached a mine to the bottom of the battleship. They were quickly captured and confessed to attaching the weapon without indicating the exact spot of its placement. But the Austrians couldn't find the mine, which exploded and sank the battleship.

The early Italian vessels were electrically propelled and had a maximum speed of 3 knots and a range of up to 10 miles. Most of these vessels and others developed during the Second World War had hydroplanes at the rear, side hydroplanes in front, and a control panel. There were typically four flotation tanks, two to the front and two aft, which were flooded or blown empty to adjust buoyancy and attitude, as is the case on a submarine. The early vessels were equipped with a compass. In some later versions, riders' seats were enclosed, and even domed cockpits were added. Most manned torpedo operations were conducted at night and during the new moon to reduce the risk of detection.

On 21 August 1940, the submarine *Iride* was attacked by three Swordfish from HMS *Eagle* while loading SLCs in the Gulf of Bomba in Libya. The submarine was sunk.

The first operational use of the SLCs was in September 1940, when eight were carried on board the submarines *Gondar* and *Scirè*, which sailed for Alexandria and Gibraltar. When *Scirè* arrived off Gibraltar, it found the harbour empty: HMS *Renown*, HMS *Ark Royal* and HMS *Sheffield* had sailed a few days earlier to help in the search for the *Bismarck*. While en route to Alexandria, the second submarine, *Gondar*, which was carrying three SLCs, passed the Mediterranean Fleet, which had left Alexandria, heading for Malta. Without viable targets, the submarine was rerouted to Tobruk. Before reaching its destination, it sighted an enemy ship and dived to avoid detection. However, it was quickly located by the destroyer HMAS *Stuart* and attacked. Two other ships joined the attack and the prolonged depth charge attack caused considerable damage to the submarine and the SLC units, which began to flood. It was forced to surface and the crew scuttled the vessel with explosive charges. All the crew but one was rescued and imprisoned.

The following month the Italian divers returned to Gibraltar but a rebreather problem prevented one team from approaching their target; another SLC sank. Finally, the third craft experienced a steering failure as it approached HMS *Barnham*. The two crew tried to drag the SLC into position but couldn't and abandoned the craft. A few hours later, the warhead exploded without causing any damage, alerting the British to this new threat. More importantly for the British, several days later, one of the scuttled SLCs was washed ashore on a nearby Spanish beach. The British wasted no time in photographing it and some of the equipment from the craft was salvaged and sent to Britain.

These setbacks led to system improvements, intensified training, and the exploration of other attack methods.

By the end of 1941, Italy was struggling to supply the Italian and German armies in North Africa. British surface ships, aircraft and submarines were so successful in blocking Italian shipping that in November 1941 only 38 per cent of materiel (29,813 of 79,208 tons) sent to Libya arrived. The Italian high command planned several measures to help remedy this situation. These included changes to convoying patterns, the use of the battle fleet to escort convoys, and SLC attacks against the British naval bases at Alexandria and Gibraltar.

On 25 March 1941, the MT were used to attack the heavy cruiser HMS *York* off Souda Bay, Crete. The motorboats were launched by the destroyers *Francesco Crispi* and *Quintino Sella* on the approaches to the bay. After negotiating the boom defences, the small craft attacked *York* and the Norwegian tanker *Pericles*. Both vessels were sunk.

Scirè returned to Gibraltar on the night of 26/27 May 1941 and launched three SLCs. One sank; the others reached merchant shipping anchored in deep water outside the military harbour but mechanical failures stopped them placing their charges.

On the night of 19 September 1941, *Scirè* launched three SLCs in the Bay of Gibraltar. Italian intelligence had reported that a battleship, an aircraft carrier and two cruisers were in the military harbour. Unfortunately for the Italians, after the earlier unsuccessful attack, the British were now more aware of this new Italian threat and better prepared to counter it. As a result, the Algeciras roadstead and the entrance to the military harbour were patrolled regularly by harbour launches, which dropped depth charges at regular intervals.

Due to the frequent patrols, the first SLC could not enter the military harbour. With dawn approaching, the crew decided to attack a merchantman and fixed their warhead to a sizeable nearby ship. Realising their intended target was an Italian-registered, captured vessel, they detached the charge and transferred it to another armed merchantman, *Durham*. The SLC crew scuttled their craft and swam to safety on the Spanish shore. The resultant explosion badly damaged *Durham* and it had to be beached by harbour tugs.

The second SLC also had trouble with the patrolling harbour defence vessels and chose a target in the roadstead. Their targets broke in half when the warhead exploded.

The third SLC managed to penetrate the heavily defended anchorage. Once inside, the crew decided they did not have the time to search for HMS *Ark Royal* and decided to attack a large tanker in the hope that a large fire would ensue and engulf the harbour. The resultant explosion broke the back of the Royal Fleet Auxiliary oiler *Denbydale*; the hoped-for inferno did not happen.

Shortly after this mission, the Italian Navy began work on *Oterra*, a tanker that had been scuttled in the harbour of Algeciras, Spain, which was within sight of Gibraltar. The Italians told the Spanish that they were cleaning the ship's trimming tanks and ensuring its neutrality. Next, the ship's bow was raised and a hatch was cut in the hull, which led to a watertight compartment. When the work was completed, the bow was submerged, which put the new compartment and the watertight hatch under the water. The Italians then told the Spanish they were moving in boiler tubes to repair the engines so the ship could be moved after the war; instead, they loaded several 22ft manned torpedoes.

Starting in early December 1942, the ship was used as a base to launch several attacks on Allied shipping, usually in open anchorages.

On 8 December 1942, three SLCs attacked British naval targets; the British harbour defence 'reacted furiously' to the attack, dropping depth charges, and three divers were killed. On 6 May and 10 June 1943, Italian SLCs from *Oterra* sank six Allied merchant ships.

A large-scale operation was conducted 26 July 1941, when MAS *451* and *452* (24.5-ton motor torpedo boats) accompanied by a 1.9-ton, two-man MTS torpedo boat, nine 1.3-ton MTM crash boats and two SLCs (carried on board an adapted motorboat called an MTL) attacked the Grand Harbour breakwater in Malta.

British radar had detected these craft well before they reached their destination, and the Italians were met with a fierce pre-warned and prepared opposition. Apart from the MTS, all the craft were lost, fifteen personnel were killed, including the unit's commander, and eighteen were captured.

A major operation was launched on 3 December 1942, when *Scirè*, carrying three SLCs, sailed from Italy. On 19 December, it made a submerged approach to Alexandria harbour, where the battleships HMS *Queen Elizabeth* and HMS *Valiant* were anchored. Italian intelligence had mistakenly reported that an aircraft carrier would be in the port.

Once in position, *Scirè* launched the SLCs. One team successfully attached a mine to the hull of *Queen Elizabeth* and another attached its mine to a large oil tanker. The final team, despite mechanical problems both with their breathing equipment and with the SLC itself, secured their charge to *Valiant*.

The resultant explosions caused considerable damage. *Queen Elizabeth* sank into the mud, although it remained upright; *Valiant* was seriously damaged but did not sink. The tanker *Sagona* was severely damaged, and a British destroyer, HMS *Jervis*, which was fuelling alongside it, also suffered severe damage. *Queen Elizabeth* took nine months to repair; Valiant six. *Jervis* was repaired and operational by the end of January. *Sagona* was towed back to England, although the repairs were not completed until 1946.

Overnight, the Italians had eliminated two of the Royal Navy's most powerful ships in the Mediterranean and at minimal cost to themselves. The attacks on Gibraltar were a nuisance. Malta showed, or so the Admiralty believed, that the human torpedoes were no real threat but this was very different. This attack represented a dramatic change in Italian fortunes and would affect British tactics from a strategic point of view during the next six months. The Italian fleet had temporarily wrested naval supremacy in the east–central Mediterranean from the Royal Navy.

2

X CRAFT DEVELOPMENT

This Italian attack prompted Winston Churchill to send a memo to the Chiefs of Staff Committee on 18 January 1942. The prime minister requested:

> Please report what is being done to emulate the exploits of the Italians in Alexandria harbour and similar methods of this kind. At the beginning of the War Colonel Jefferies had a number of bright ideas on the subject, which received very little encouragement.

In late March 1942, New Zealander Captain William Fell, who had been the commanding officer of the infantry assault ship HMS *Prince Charles*, was visiting the headquarters of the Submarine Service at Northway House, London. He had served on submarines since 1918, but during the war he had been posted to Q ships, followed by various positions in Combined Operations, and now he was looking for a new appointment. While there he met Max Horton, Flag Office Submarines, who asked him if he would like to return to submarines. Fell replied in the affirmative and Horton said, 'go away and build me a human torpedo'. Fell was briefed about the Italian attack on the fleet in Alexandria and told that two of the human torpedoes had been salvaged and were being transported to HMS *Dolphin*.

That evening Fell drove to *Dolphin* and the following day he met Captain Reginald Drake DSO, Commander of the 5th Submarine Flotilla. Together with the base's chief engineer, Commander Stanley Terry, they started to develop the British human torpedo, which they named Cassidy. At this stage, the actual Italian vessels had not arrived in Britain, so they had to use photographs. They started with a 20ft-long wooden log, 2ft in diameter. The front was rounded and the back tapered. A 2-gallon tank was attached to each end and pipes, valves and a pump connected so that water could be flooded or pumped out from the tanks or water could be transferred from one to the other. In the middle was another tank that could be filled from the sea and emptied by being blown by compressed air. A lead keel was fitted because the prototype vessel tended to roll over when lowered into the water. Once the craft was stable, hydroplanes and a rudder were fitted; a joystick controlled these.

During this period, another team member, Commander Geoffrey Sladen DSO, was at the Siebe Gorman factory at Tolworth, developing a diving suit for the two-man crews of the human torpedoes.

This resulted in a very heavy, cumbersome dry suit, which became known as the 'Sladen Suit'. The diver put on the suit by climbing through a wide rubber tube at the navel; once 'inside', the tube was folded and tied off before the diver entered the water. The first version had two small, glazed eyepieces. It was redesigned with a single oval flip-up viewport so the wearer could use binoculars. The divers that had to wear it referred to the suit as 'Clammy Death'.

Recruiting the men that would crew the vessels started in May 1942. Initially, Sladen and Fell approached newly joined RNVR officers; out of a thousand men, 200 showed an interest in this new project, which had become known as special submarine-related service. Out of these 200, eleven made it through the first screening. They were then posted to HMS *Dolphin* in Gosport, Hampshire, to join the 'Experimental Submarine Flotilla', where they joined a small number of junior rates.

Initial training took place, wearing standard diving suits and hard helmets with air supplied from a surface compressor because the Sladen diving suit was not yet available. There were also many short trips on submarines.

Once the men had become used to working underwater, they were introduced to the Davis Submerged Escape Apparatus (DSEA). This was an early type of oxygen rebreather, invented in 1910 by Sir Robert Davis, who was head of Siebe Gorman & Co. Ltd. It was an improved version of the Fleuss system developed by Henry Albert Fleuss, who was a pioneering diving engineer and Master Diver for Siebe, Gorman & Co. The Royal Navy finally adopted the DSEA after further development by Davis in 1927.

It consisted of a rubber breathing/buoyancy bag containing a canister of barium hydroxide, which removed carbon dioxide from the exhaled CO_2. In the bag was a small steel cylinder that held about 56 litres of oxygen at a pressure of 120 bar.

A flexible corrugated tube connected a mouthpiece to the canister of CO_2 absorbent inside the breathing bag. When required, oxygen could be bled from the cylinder into the bag by opening a control valve. In addition, a non-return valve was fitted in the bag that allowed air, which would expand as the wearer rose towards the surface and the water pressure decreased, to escape. On reaching the surface, the valve was closed, making the DSEA a life preserver.

The DSEA used by the Royal Navy had an extra emergency buoyancy bag attached to the front of the main breathing/buoyancy bag. This was to keep the wearer afloat on surfacing. This emergency bag was inflated by opening a small steel oxygen cylinder. As users could only breathe through their mouths, a nose clip was supplied, as were goggles. Although this equipment was principally used as emergency escape equipment for submarine crews, it was soon also used for diving, being a handy shallow-water diving apparatus with a thirty-minute endurance, and as an industrial breathing set.

Although several recruits failed the initial training and were returned to their units, new volunteers kept the fledgling unit's strength at about twenty men. To begin with, the men didn't know what they were being trained for; many assumed they were training to be divers who would be deployed from submarines. However, once they were told of their intended purpose, Fell was impressed with their excitement and enthusiasm, whose 'one idea was to get at the enemy'.

After several weeks they got a prototype, although it had no form of propulsion. The team used it in Horsea Lake in Portsmouth, where it was towed by a boat so the crews could practise diving, surfacing and working while wearing the Sladen Suit. By 1 June, the fledgling flotilla had grown to a complement of fifty-seven men.

One major problem that became apparent during the initial training was the very dangerous blackouts caused by breathing oxygen at depth for prolonged periods. One young officer lost his life, despite strict and robust safety precautions.

The crews referred to this as 'lips': the first sign of oxygen toxicity was an uncontrollable twitching of the lips. Breathing pure O_2 at depths greater than 13ft can result in several health risks such as dizziness, convulsions, blurry sight and, in the worst cases, drowning. A programme was started at the Siebe Gorman works at Tolworth, where approximately 1,000 test dives were carried out to assess the effects of oxygen under pressure and formulate a safe practice.

Horsea Lake hardly provided a realistic training environment, so Fell relocated his men to Loch Eriscort in the Outer Hebrides. HMS *Titania* arrived on 12 June 1942. The loch provided the isolation and security required, although the trainees had to build a jetty and huts ashore.

Unfortunately, this location also provided extremely bad weather, which severely limited the amount of available training time. To remedy this, the operation was moved to Loch a'Choire on the western side of Loch Linnhe on the west coast of Scotland; it also provided an ideal anchorage for the squadron's depot ship, HMS *Titania*.

By late 1942, the British had developed their own manned torpedoes. Stodhert and Pitts of Bath built the prototypes and they became known as chariots. There were two versions, 20ft and 30ft long, capable of speeds of 2.5 and 4.5 knots respectively, and each carried two men. They were capable of deployments of five and five and a half hours. In late September 1942, Fell was confident enough in the efficiency of his men to tell Flag Officer Submarines that from 9 November 1942 four crews would be ready for operations.

Irrespective of the prime minister's prompt, the Navy had realised early in the war that the major units of the German and Italian navies posed a considerable threat to the British lines of communication. Even at anchor in their remote, highly defended bases, they required the redeployment of many British warships. It became apparent that the enemy ships were probably more vulnerable to underwater attack when in harbour than when they were at sea. To this end, the production of a midget submarine (the X craft) was already under way. The prime minister's memo led to a stopgap solution until the X craft were ready for deployment. This intermediate solution was the human torpedo, where unlike the midget submarine, the two crew rode the 'chariot' wearing diving suits and breathing apparatus. The British had salvaged an Italian human torpedo after an attack on Gibraltar, forming the basis of the British design. Once enough volunteers had completed initial training at Fort Blockhouse in Gosport, they joined HMS *Titania*, which sailed for Loch Erisort. This was to become the charioteers' base, known as Port D.

Commander Godfrey Herbert RN (Retired) devised the first British 'midget submarine' in 1909. His one-man guided torpedo, which he christened 'Devastator', was intended to 'provide a means for propelling against an enemy ship or other target a large quantity of high explosive, and of effecting this with great economy of material and personnel' – or so the commander's patent would have us believe. Despite this, the idea was turned down by the

Admiralty. In 1915 Robert H. Davis of Siebe, Gorman and Co. Ltd patented a design for a three-man submarine that incorporated a floodable compartment so crew members could leave and enter the submarine underwater. When he was captain of Submarines at Fort Blockhouse in Gosport, Max Horton proposed three designs, none of which were approved by the Admiralty. Winston Churchill was the First Lord of the Admiralty at this time.

Commander Cromwell Varley presented the proposal, which was eventually developed into the X craft. DSO RN (Retired) during the interwar years. He had commanded the submarine *L1* during the First World War and had been retired from the service in 1920. His initial plans for the midget submarine were for a craft of approximately 26ft to be crewed by two men. However, by 1940 he had increased the length of the submarine to 50ft – the extra space accommodated an escape compartment similar to the one proposed by Robert Davis some years earlier. The Admiralty, who were becoming increasingly concerned by German capital ships harbouring in Norwegian fjords, accepted this idea. Against this background, approval was given to construct the first X craft, *X3*. (*X1* had been an experimental submarine and *X2* was a captured submarine). The miniature submarine was built at Commander Varley's boatyard, Varley Marine Ltd. So important was the need for secrecy that it was launched at 11 p.m. on Sunday, 15 March 1942, into the River Hamble, just off Southampton Water. The crew, who had been standing by her for the last six months of its build, were Lieutenant W.G. Meeke, DSC RN, Lieutenant D. Cameron RNR and Chief Engine Room Artificer Richardson. These three men had served in submarines for several years, so they were not unfamiliar with their new charge. *X3* was a real submarine: forward was the control room, which housed the steering and depth-keeping controls, and the periscopes and navigational equipment; aft of this was the escape compartment known as the 'wet and dry', or 'W' and 'D'; further aft were the battery compartment and machinery spaces.

X3 was, in nearly every respect, a submarine but in miniature. The major difference between *X3* and its larger sisters was the armament. The X craft did not have torpedo tubes; instead, it carried two crescent-shaped explosive charges, known as the side cargo, each containing 2 tons of amatol explosive, which were intended to be dropped under the enemy target. During the next few months, *X3* carried out numerous trials until 26 August 1942, when the submarine was sailed to Southampton. There it was loaded onto a special railway truck and began the journey north to Faslane. Then, with the help of one of the military port's large cranes, it was offloaded into the Gareloch and sailed to its new base at Port Bannatyne on the Isle of Bute.

The following six X craft (*X5–X10*) were built at Barrow-in-Furnace by Vickers and commissioned from December 1942 to February 1943. They had slightly different internal layouts to *X3*. All were lost during the attack on *Tirpitz*. A further class of midget submarines, XE craft, were built and completed between December 1943 and January 1945. These craft had the luxury of air conditioning and were intended for service in the Far East. They were built in the north of England by engineering firms more used to manufacturing farming and mining equipment. Six submarines were constructed purely for training purposes and these were known as XT craft. The limited range of the small submarines made it necessary to tow them to the target area with a conventional submarine. Because of the cramped and unpleasant conditions and the long tow times (the journey to Norway took eight days), the X craft had two crews: a transit crew and an operational crew. During the outward voyage, the transit crew would prepare the boat for the operational crew, who travelled in the comparative luxury of the towing submarine. Once in the target area, the crews would change over and the operational crew would carry out the attack.

BUILDING X CRAFT SUBMARINES

A memoir by James Henry Weatherburn (ex-Vickers Ltd):

> Having been asked to produce an account of my experiences associated with X Craft production, ordered from Vickers Barrow for the Royal Navy, during the Second World War, I will record what I remember but it was some years ago and I have worked on a great many submarines since.
>
> My first glimpse of a Midget Submarine was about August 1942 when my then foreman in the Submarine Dock sent me to have my photograph taken to form a special pass to enter a secure area set in the North Shop (formerly the Gun Shop and now the Nuclear Build Shop) to work on a special project. I had no idea what the project was and doubted whether anyone else working in my area did either.
>
> I had just entered my final apprenticeship year (5th) when I arrived in this area to work for the same foreman I had been working for in the

Sub Dock and was pleased he had included me in his new team. To my surprise I was given to work on my own as a journeyman and was taken to what I recognised as a unit of a small submarine because I was to fit a rudder and a hydroplane to this unit with its operating rod assembly to the three glands to inside the unit. Other units arrived in the workshop and then could be seen the elements of a mini-submarine by then known as X Craft. Each complete submarine consisted of three units – a Bow Unit, Control Room Unit and Tail End Unit each with a joint ring flange at its open ends with probably 40 bolts to join the units together. This design was good for build as outfit could be completed as far as possible with accessible open ends.

The bow unit had a forward trim tank and other small tanks under a battery of, I think, about 112 batteries which, when installed reached up to above halfway and then a wooden platform above it sufficient for a man to carry out battery maintenance and also use it for sleeping in service.

The next unit was mainly the control unit which had a wet and dry compartment at its forward end, enclosed by two bulkheads, each with a circular bulkhead door: the forward one for entrance to the battery tank and the aft one for entry or exit to the control room. This W & D compartment also had a hatch above it and was to enable a diver to exit and re-enter the submarine for the purpose of cutting submarine nets and also laying charges under an enemy ship as well as undercover coastal work. This unit also had a periscope observation blip on the hull, two very important hull castings on each side which operated the release of side explosive charges and a pressure hull hatch aft. This was also a blessing for access when the units were joined.

The aft unit contained a rudder and aft hydroplane, aft trim tank, fuel and water tanks, a Keith Blackman motor coupled directly to the propulsion shaft, a Gardner diesel engine – similar to the London bus engines with a clutch between to disengage the engine. All connecting systems to the control room unit required to be exactly positioned so that the complete spare tail end unit could replace any of the six Units.

The operating equipment was small and the valves were neatly designed. The trim control was a bit like a car gearbox where you had crossed slots. Pushing away the toggle discharged water out, pulling the toggle towards you fed water into the tanks, go left in the slots – Aft Trim to for'd trim and right in the slot – for'd trim to aft trim. The engine control was right next to this and, I think, the aft hydroplane control. You could not stand up in the control room so always had dipped head and shoulders.

Dawned the day when the three units were joined together with a thick dexine joint and numerous bolts and then what marvellous tiny submarines we had on display. However this brought its problems with production because if you had eight or ten people inside you could hardly move and men completed their tasks working on top of each other, all in very good spirits as I remember. Men sat at the forward hatch waiting to get in as men left the aft hatch.

Eventually we had the first complete boat lifted onto a railway wagon and covered with a tarpaulin. I noticed a name had been painted on the bow – I think it was *Shrimp*. Later one was called *Platypus* and one with a Greek or Latin name beginning with X [probably *Xerxes* – ed.]. It was also at this time that I saw two young Sub Lieutenants with the boss looking around the outside of the boat who I later recognised as Cameron and Place. Arriving for work next morning, no railway wagon, no submarine, it had disappeared over night. The train would have been escorted, I believe, to Faslane by crew members and then on to Rothesay.

This first boat was probably *X5* and, as *X6* was nearing completion, my Boss asked me to accompany one of his Fitters (Bill Moscrop) to Scotland to carry out a modification to the clutch on the submarine recently dispatched. We caught the 5.30 a.m. train out of Barrow and travelled to Wemyss Bay, where a ferry plies to Rothesay. It was six thirty and pitch black when we arrived at Rothesay, where a naval truck was waiting for us. They drove us to Port Balantyne [*sic*, Bannatyne] and immediately transferred to a waiting liberty boat. They ploughed through the dark for about two hours to what was, I later discovered, Loch Striven. A series of wooden huts were built here and a jetty, alongside which was our *X5*.

Bill got stuck in right way on the job and I think we finished about six o'clock in the morning and, boy, did the bacon and egg go down well which they provided but the huts and bunks were a bit rough and the boys were being pushed hard in their training.

When dawn broke I could then see the set up here was a large mansion set back from the Loch which was where the officers were accommodated but otherwise it was the end of nowhere. There were the two-man torpedo riders tied up at the jetty but pretty quick the riders were on them and off up the Loch and diving. We had walked part way up the hillside to watch our *X5* move out and dive and then we were taken back to Port Bannatyne. I think I slept on the boat and later on the train. This had a profound effect on myself because here you see the real frontline of the war, these young men, none much older than I, risking their lives in a very positive way, and we knew from conversation that several lads had already died, what brave lads. I, myself, had seen in Barrow ships coming in damaged – bows blown off, we produced every type of weapon and we had a blitz but the real coal face was not experienced.

It was not long after returning that I was told to accompany another fitter (Bill Kelly) to Rothesay to carry out Mods on *X5*, *X6* and *X7*, which had left Barrow to join the Navy. This time the journey from Barrow was the same. We were escorted into HMS *Varbel* at Port Bannatyne, which had been the Hydro Hotel but taken over by the Navy, and was now a ship. We gathered that the three subs were tied up by a steam barge alongside the Jetty. As we entered HMS *Varbel*, Bill Kelly was spoken to by a couple of ERAs [engine room artificers]. We were marched up the steps and into a large room full of officers and Captain Ingram sat in the middle. He welcomed us and said that everything was ready and laid on to start work right away.

I didn't know what Bill was thinking, I was only the helper, but he said we wouldn't start right away but would start in the morning. He told the Captain we had travelled for 14 hrs and because there were three boats we should get some sleep and then work right through. I think the Captain was livid and told an officer to get on with one boat himself. It didn't make sense what Bill said but we then went down the pub, where quite a few of the crews were having a night ashore. It seems they had whispered to Bill not to work or they would be kept on standby. To be honest they had been stuck up at Loch Striven for weeks on end and wanted a break. They talked about their families and their futures and I was more moved when, later on, the *Tirpitz* raid was announced and you realised they must have known they would not get back home.

We started work early next morning and worked right through till dawn the next day and had to do all three boats. I remember lying on the deck of the steam barge as dawn broke and the deck was lovely and warm. It took us so long because of the access lying full length on your tummy with pipes and brackets in your back. We could only take it in turns on one boat at a time.

The six boats were completed by January 1943, so were all completed in about 5 months. I read, after the war, an article which said that they were required mainly for a raid on *Tirpitz* and they had two windows – March and September. However they were not ready and fully trained by March so missed that earlier time.

I returned to larger submarine building for a little while and then, which must have been around June/July 1943. My same foreman (Ted Fleming) had to take a team (maybe six or eight men) to Port Bannatyne for modifications to all six boats *X5* to *X10*. When we arrived all six boats were nose to tail on a floating dock and their depot ship HMS *Bonaventure* was anchored nearby in Rothesay Bay. We carried out many modifications which I cannot remember now but I think it was extra buoyancy forward. We were there two or three weeks and had digs ashore with some old ladies, poor dears, who tried to feed us but rations were difficult in war time. However we used to feed at lunchtime and tea time on *Bonaventure*. Boy, what a treat, we had white bread, baked on board, which we hadn't seen since 1940 and lashings of real butter, bacon, eggs, sausage and home-cooked puddings etc. There was no rationing there and quite right too, these were the real heroes of the war.

3

12TH FLOTILLA OPERATIONS

These new submarines were based at HMS *Varbel*, the former Kyles Hydropathic Hotel at Port Bannatyne, on the hill overlooking Kames Bay on the Isle of Bute. Between September 1942 and May 1945, the hotel was requisitioned by the Royal Navy as the headquarters of the 12th Submarine Flotilla and named after Commander Varley, the X craft designer, and Commander Bell, the flotilla training commander. It was commissioned on 11 September 1942, under the command of Commander David Ingram, DSC RN, a former submarine captain. The long, three-storey building housed the administration offices and messes of the fledgling submarine flotilla. Near the head of Loch Striven on the eastern shore, the flotilla had another shore base: a shooting lodge called Ardtaraig House that was commissioned on 21 November 1942 as *Varbel II*. The midget submarines *X3* and *X4* were based there for training purposes. The lodge had an outside swimming pool, which the X craft crews used for training, and there was also a complete 'wet and dry' section that could be lowered into the loch and provided the crews with a further training facility. When the newly joined X craft crews first came to Scotland, they spent the first week at the Hydro completing basic instruction. Then they spent three weeks at *Varbel ºII* training on the X craft, where they practised diving and net cutting – all the facets of their new, precarious trade. The submarines were maintained at Ardmaleish Boatyard, a few miles to the north of Port Bannatyne.

The hotel was situated on the side of the hill overlooking Port Bannatyne and the pier where the submarines would berth. Two roads led from the pier to the Hydro. One went there directly and the other passed by the Royal Hotel, which was favoured by the off-duty submarine crews. The landlord, Peter McDougal, and the barman, Willie Friel, always welcomed the midget submarine crews. Perhaps more importantly, the Royal's beer was said to be better than the Hydro's – even though the senior rates barman, PO Capling, an admiralty messenger, endeavoured to improve the Hydro's drawing power and appeal by operating somewhat flexible bar hours! The ship's bell hung in the Hydro's spectacular entrance foyer; this bell was 'second-hand'. It originally belonged to the merchant ship SS *London Merchant*. It was renamed *Politician* and on 5 February 1941 it ran aground on the island of Eriskay. The ship carried a large quantity of whisky, which the islanders salvaged. This incident was later immortalised in Sir Compton Mackenzie's *Whisky Galore*.

The X craft crews prided themselves on returning to base shaved and with, more or less, clean hands. Regardless of this, it still took three or four days and several baths in the Hydro for the men to feel clean enough to ask one of the Wrens for a dance.

The young men of the 12th Flotilla had volunteered for hazardous service and tragically there were a series of fatalities during their training. Sub-Lieutenant David Locke was killed by oxygen poisoning on 31 May 1942 while trying to cut through a practice submarine net in Loch Striven. The next month, *X4* was sailing, on the surface, off the Isle of Bute when Sub-Lieutenant Morgan Thomas was swept out of the wet and dry compartment's open hatch. His body was never found. On 4 November 1942, during the afternoon, *X3* began to dive in Loch Striven when a valve broke and the submarine began flooding. Attempts to shut the valve failed and the craft sank stern first at a very steep angle and settled on the seabed at 120ft. Chlorine gas was released when sea water got into the batteries. The crew managed to put on their escape sets and get out through the hatch in the wet and dry compartment. Fortunately, all three men survived and *X3* was later salvaged and repaired.

From this very imposing base, men of the 12th Submarine Flotilla waged their war against Germany with their midget submarines and human torpedoes. The X craft were originally designed to attack the German battleship *Tirpitz*, which was berthed in a Norwegian fjord, where the Germans presumed she would be safe from all normal means of attack.

On 26 October 1942, Operation Title saw the charioteers, operating from the Norwegian fishing vessel *Arthur* under the command of Leif Larsen of the Royal Norwegian Navy's Special Service Unit, come practically within sight of their target when the mission had to be aborted. A severe north-easterly gale buffeted the small trawler and this placed an immense strain on the towing cable. The cable snapped, both chariots were lost and *Arthur*'s propeller was damaged. The boat was scuttled and the men all set off on foot to try to escape overland to neutral Sweden. They split into two groups and for five days and nights walked eastwards towards Sweden. One group made it across the border undetected. The Germans found the second group close to the Swedish border. A gunfight ensued and Able Seaman Bob Evans was shot in the leg. As he was unable to walk and the alarm had been raised, the decision was taken to dress his wound and leave him to be taken into custody by the Germans. He was in Royal Naval uniform, so he would be entitled to the protections of the Geneva Convention. The remaining four men successfully crossed the border. Unfortunately, Evans was shot by the Nazis under Hitler's Kommandobefehl (Commanding Order), which stated all raiders and saboteurs should be shot regardless of uniform or military status.

The flotilla made a second attempt on 22 September 1943 (Operation Source), when six X craft were deployed to attack the warship. *X5*, *X6* and *X7* were directed to attack *Tirpitz*, *X8* the *Lützow*, and *X9* and *X10* the *Scharnhorst*. On 11 September, six conventional submarines, each with an X craft in tow, sailed for Altenfjord. During transit, *X9* was lost at sea with all hands. *X8* suffered a series of mechanical failures culminating in the premature explosion of one of its side cargoes, leaving no choice but to scuttle it. As a result, the *Lützow* attack was aborted.

On 20 September, the remaining four X craft slipped their tows and started their attack. Unfortunately, *X10* began experiencing mechanical problems; its commanding officer, not wanting to compromise the whole mission, aborted his attack and returned to the rendezvous point. Although he didn't know at the time, his target, *Scharnhorst*, had already sailed from its anchorage.

The three craft that were to attack *Tirpitz* would spend the night of 21 September at Brattholm Island, recharging their batteries, and then carry out their attack the following day.

Tirpitz was anchored in the Kaafjord, a branch of the Altenfjord, protected by two anti-submarine nets. Just before dawn, the outer anti-submarine net was opened to let a coaster through. *X6*, commanded by Lieutenant Donald Cameron, followed the coaster and managed to slip through the net before it was closed. The submarine hit an uncharted rock and briefly surfaced, at which point she was spotted by guards. At this stage, it also had a faulty periscope and gyroscope. Despite this, Cameron managed to get *X6* under *Tirpitz* and release his charges. Knowing escape was impossible, he surfaced

X6 and scuttled the craft; Cameron and his crew surrendered and were taken prisoner. Unfortunately, they were taken on board *Tirpitz*. *X7*, commanded by Lieutenant Basil Godfrey Place, despite getting entangled in the nets, also managed to lay its charges. While making its escape, it became entangled in the net again.

At 8.12 a.m. the charges detonated and the shockwaves from the explosion freed *X7*, but it was damaged too severely to leave the scene and Place quickly surfaced. Two crew members were unable to escape before *X7* sank. The blast also caused *X5*, commanded by Lieutenant Henty-Creer, to surface; it was immediately attacked by a combination of gunfire and depth charge from *Tirpitz* and sank. Although it was thought that it never had a chance to release its charges, this was never proven conclusively and there is a possibility it managed to plant them before being sunk.

The charges caused considerable damage to the ship. The first exploded abreast of Caesar turret, with a second detonating approximately 150ft off the port bow. These ruptured a fuel tank, flooded compartments, damaged shell-plating, severely damaged the keel, and bulkheads in the double bottom buckled. The ship flooded, causing a 2-degree list, which had to be balanced by counter-flooding compartments on the starboard side. Electric supplies throughout the ship were badly disrupted. One gun turret was unseated from its bearings and could not be trained, rendering it useless. The ship's two Arado Ar 196 floatplanes were completely destroyed.

Cameron and Place received the Victoria Cross for their roles in Operation Source. Lorimer was one of three who received DSOs. There is a monument honouring the crews of *X5*, *X6* and *X7* in the churchyard at Kaafjord.

It took a monumental effort to make *Tirpitz* seaworthy again. The repair ship *Neumark* carried out the repairs, which lasted until 2 April 1944; full-speed trials were scheduled for the following day in Altafjord. Repeated attacks by RAF and Fleet Air Arm aircraft delayed the final sailing. On the day it was due to sail, 15 September 1944, twenty-three RAF Lancaster bombers of 9 and 617 Squadrons flew from an airfield at Yagodnik in northern Russia and attacked the battleship in Operation Paravane. Mountains screened the low-flying planes from radar, which caused the Germans to be slow in deploying the smokescreen and presenting the Lancasters carrying Tallboy bombs with a relatively clear target. One bomb went through the ship's bow and exploded on the seabed. The bow flooded, causing the ship to sink by the bow. Also, its speed was limited to 8–10 knots. There was also severe damage to fire-control equipment. All the Lancasters returned safely to the Russian airfield.

The Germans decided that it would not be possible to make the ship completely seaworthy again. It was moved further south to Tromsø on 15 October 1944, where it was intended to use it as a heavy artillery battery. On 12 November 1944, the RAF launched the final attack on *Tirpitz*, Operation Catechi. Thirty-two Lancaster bombers dropped twenty-nine Tallboys, achieving two direct hits and a near miss. Several other near misses destroyed a sandbank the Germans had built up to keep the ship from capsizing. Fifteen minutes after the bombers attacked at 9.35 a.m., *Tirpitz* was listing 60 degrees to port. A large explosion blasted one of the main turrets into the air. By 10 a.m., the ship had capsized.

In January 1943, five chariots were launched near Palermo, Sicily. One lost its limpet mines and the equipment needed to secure the warhead to the target when a large wave swept over it. A second chariot was also damaged approaching the target area. The remaining three chariots continued into Palermo harbour, where they sank the Italian cruiser *Ulpio Traiano* and badly damaged the converted liner *Viminale*. All the chariots were lost in the raid and one British submarine was also sunk. One charioteer lost his life in the attack, and seven others were captured.

That same month, two British chariots were deployed to Tripoli in North Africa and used to help prevent blockading ships from being sunk at the harbour mouth.

X craft struck at other targets, including the attack on the floating dock at Bergen, which restricted U-boat repairs and reduced the amount of U-boat activity off the coast of Norway. *X24* sailed to attack the floating dock in Bergen on 9 April 1944. Although the attempt was not entirely successful, a large German merchant ship was sunk and considerable damage was done to the dock facilities. *X24* returned to Bergen on 3 September 1944 and completed the mission, sinking the dock.

Operation Heckle saw *X24* attack the floating dock at Laksvaag, Norway. It was being towed across the North Sea by HM Submarine *Spectre* when Sub-Lieutenant Purdy was unfortunately lost overboard.

British chariots attacked Italian shipping in Sardinia and Sicily in Operation Principle in January 1943. The chariots were transported by three submarines: HM Sub *P311*, HM Sub *Trooper* and HM Sub *Thunderbolt*. Eleven charioteers died during this operation.

Sub-Lieutenant Page on the casing of *X10*. (*National Museum of the Royal Navy*)

A party of charioteers was deployed to Malta, where it joined Captain G.W.G. Simpson and his famous 10th Submarine Flotilla. From its base on Manoel Island it was transported by specially adapted submarines (*Thunderbolt*, *Trooper* and *P311*) and carried out attacks on the harbour at Tripoli and the north Sicilian port of Palermo. It also carried out beach reconnaissance before the Allied landings on Sicily.

HMS *X20* spent four days off the Normandy coast conducting periscope reconnaissance of the shoreline and echo soundings during the daytime. Each night, *X20* approached the beach and two divers swam ashore and collected beach samples, which were placed in condoms. The divers went ashore on two nights to survey the beaches at Vierville-sur-Mer, St-Laurent-sur-Mer and Colleville-sur-Mer; the location of the American Omaha Beach operation on D-Day.

On 21 June 1944, a joint British–Italian operation was mounted against shipping in La Spezia harbour in north-west Italy. The chariots were carried on board a Motor Torpedo Boat (MTB) and the cruiser *Bolzano* was sunk.

Operation Gambit was part of Operation Neptune, the landing phase of the invasion of northern France (Overlord) during the Second World War. Gambit involved two X-class submarines (*X20* and *X23*), marking the ends of the British–Canadian invasion beaches. Using navigation lights and flags, the submarines indicated the western and eastern limits of Sword and Juno

Beaches. The midget submarines arrived on 4 June and, due to the delay caused by bad weather, remained in position until 4.30 a.m. on 6 June (D-Day). They then surfaced and erected the navigational aids, an 18ft (5.5m) telescopic mast with a light shining seaward, a radio beacon and an echo sounder, tapping out a message for the minelayers approaching Sword and Juno. The submarines were at some risk of damage due to friendly fire and to avoid this, Lieutenant George Honour, the captain of *X23*, flew a White Ensign of the size usually used by capital ships. A similar operation had been offered to the Americans to mark their beaches but this was declined.

On 6 March 1945, *XE11*, while carrying out a calibration exercise in Loch Striven, collided with a boom defence vessel after drifting out of its exercise area. Three of the crew were lost in the accident but two managed to escape. The boat was later salvaged.

I had the pleasure of meeting Bill Morrison on a number of occasions over several years. On a few occasions, he told me about his time on *XE11*. He was the submarine's first lieutenant; the collision happened just eleven days before his 21st birthday.

The submarine had been launched at Faslane on 19 February 1945 and sailed to HMS *Varbel* at Port Bannatyne. On the morning of 6 March, *XE11* set sail from Port Bannatyne for a routine training exercise in Loch Striven. It was carrying a crew of five. In command was Lieutenant Aubert 'Eustace' Staples, a 24-year-old Rhodesian naval officer of the South African Naval Forces who had come to Britain to join the war effort. Second in command was Bill, only 20 years old, a sub-lieutenant of the Royal Navy Volunteer Reserve. Also on board were Able Seaman 'Scouse' Carroll, Engine Room Artificer (ERA) Les Swatton and Stoker Higgins, who had married three weeks before the accident.

The purpose of the exercise was to calibrate some of the instruments. This would require the submarine to dive to 100ft and then gradually rise in 10ft steps, stopping at each increment to check and calibrate instruments. During the course of the exercise, *XE11* had drifted out of its designated exercise area, although the crew were unaware of this. The boom defence vessel *Norina* was laying a line of buoys in Loch Striven that morning.

At 11.15 a.m., *Norina* was lying stationary with its engines stopped; it drew a depth of 9ft and now the submarine was directly underneath it. The boom defence vessel would have been undetectable to the crew of *XE11* and they had no way of knowing how far outside their allotted area they were.

The X craft had completed its final calibration check at 10ft and was about to surface. Lieutenant Staples told Bill to go into the wet and dry compartment so he could open the hatch once the submarine broke the surface.

At 11.20 a.m., a large crash was heard at the for'd end of the submarine as they started to surface. The submarine heeled over and then righted herself. Lieutenant Staples told Bill to check the battery compartment for damage. The submarine slid along the hull of *Norina*, producing a dreadful scraping sound.

Unfortunately, at this very moment, the ship started her engines and moved ahead, and her propeller cut into the submarine's pressure hull; the submarine

Four officers on the casing of *X21* in their Ursula suits. (*National Museum of the Royal Navy*)

immediately began to flood. Lieutenant Staples ordered 'full ahead' and the planes to 'hard to rise' in an attempt to drive the craft to the surface but as the control room flooded, it began to sink stern first.

In the wet and dry compartment, Bill was desperately trying to open the hatch so the crew would have a means of escape but as the submarine sank, the water pressure increased, making it more difficult to force the hatch open.

Lieutenant Staples ordered ERA Les Swatton into the wet and dry to help Bill open the hatch. He then handed out the DSEA vests to everyone. Unfortunately, the situation worsened when the incoming water shorted the electrical system and the lights failed.

XEII settled, luckily upright, on the loch bottom at a depth of 180ft. Due to the flooding, the pressure in the submarine quickly equalised and the wet and dry hatch blew open.

As the remaining air escaped, Bill and Les were jammed together in the hatch. Bill managed to work his way back into the wet and dry compartment and was able to push Les through the hatch towards the surface. Bill remembers reaching back into the control room, trying to get hold of any of his crewmates; regrettably, he couldn't reach anybody and in the darkness, he passed out. Somehow his unconscious body was swept up through the open wet and dry hatch and his DSEA carried him up to the surface.

The crew of *Norina* quickly rescued the two men. Bill was unconscious when he was lifted aboard the ship and was given mouth-to-mouth resuscitation. Both men were transferred to medical facilities at HMS *Varbel*. Thirty-seven years later, Bill discovered that he had fractured his neck while trying to force the wet and dry hatch open.

Unfortunately, they were the only members of the midget submarine to escape. The close-knit community of HMS *Varbel* were stunned; in the space of just a few hours, they had lost three shipmates, three friends: Lieutenant Staples, Able Seaman Carrol and Stoker Higgins did not return alive.

Over the following days, helmeted divers salvaged *XEII*. The salvaged submarine was moved to the boatyard at Ardmaleish, where the bodies were removed and a thorough technical examination was carried out.

Bill Morrison and Les Swatton held the record for the deepest escape from a submarine for many years. They were included in the Guinness Book of Records as having made the deepest unaided ascent from a sunken submarine. This lasted until July 1987, when approximately twenty-five instructors from several countries, including Australia, Norway, Canada, Turkey and Israel,

X24 with a crew member on the casing and flying the Jolly Roger. (*National Museum of the Royal Navy*)

An X craft under construction – officers on the casing. (*National Museum of the Royal Navy*)

carried out a deep escape exercise from HMS *Otus* in a Norwegian fjord 25 miles north of Bergen. They escaped, using the one-man tower, from 600ft.

The bodies of the three young men who had died were buried side by side in Rothsay cemetery beside what is now the United Church of Bute. During the Scottish Submarine Association Branch Dunoon Weekend a small service is held at these graves.

* * *

When the war in Europe was almost over, the first six XE craft were deployed to the Far East with their depot ship *Bonaventure* and formed the 14th Flotilla, sailing from Subic Bay in the Philippines.

In August 1945, HMS *XE1* and *XE3* carried out a joint attack on Japanese heavy cruisers anchored in Singapore harbour. *XE3* was tasked with attacking *Takao*, while *XE1* was to attack *Myōkō*. *XE3* approached *Takao* through the Straits of Johor and the various harbour defences. That took eleven hours and a further two hours were needed to locate the camouflaged target. The Japanese defenders failed to spot the vessel and the submarine successfully fixed limpet mines and dropped its two 2-ton side charges. *XE3* safely returned to HMS *Stygian*, the escort submarine. Meanwhile, the crew of *XE1* failed to locate their assigned target and decided to attach *Takao*, despite being aware that the explosives already laid by *XE3* could explode. *XE1* laid its charges under the ship before making a successful escape.

Although *Takao* was already damaged and not seaworthy, it was severely damaged in this attack and never sailed again. Lieutenant Ian Edward Fraser RNR and Leading Seaman James Joseph Magennis were awarded VCs for their part in the attack. Sub-Lieutenant William James Lanyon Smith, RNZNVR, who was at the controls of *XE3*, received the DSO. ERA Third Class Charles Alfred Reed, who was at the wheel, received the Conspicuous Gallantry Medal (CGM). *XE1*'s commanding officer, Lieutenant John Elliott Smart RNVR, received the DSO; Sub-Lieutenant Harold Edwin Harper, RNVR, received the DSC; and ERA Fourth Class Henry James Fishleigh and Leading Seaman Walter Henry Arthur Pomeroy received DSMs. In addition, ERA Fourth Class Albert Nairn, Acting Leading Stoker Jack Gordan Robinson and Able Seaman Ernest Raymond Dee were mentioned in despatches for their part in bringing the two midget submarines from harbour to the point where the crews that took part in the attack took over.

The 14th Flotilla also cut the seabed telephone cables that linked Japanese-occupied Saigon with Singapore and Hong Kong. These operations, carried out in July 1945, were intended to make the Japanese use radio, thereby rendering themselves open to message interception.

On the first mission, Operation Sabre, *XE4* attempted to cut the Hong Kong to Saigon telephone cable and was towed to the target area (the Mekong Delta) by HMS *Spearhead*. *XE4* searched for the two telephone cables using a towed grapnel. When it found the first cable, it was cut by Sub-Lieutenant K.M. Briggs using the net/cable cutter. The second cable was cut by the second diver, Sub-Lieutenant A. Bergius. Two divers were carried because a diver should not spend more than twenty minutes at depths of over 33ft and no more than ten minutes at over 40ft. *XE4* safely returned to *Spearhead* and docked at Labuan on 3 August 1945. The second cable, the Hong Kong to Singapore telephone line, was cut in Operation Foil by *XE5*, after being towed to the area by HMS *Selene*. Operating close inshore near Lamma Island, *XE5*'s divers had to work in appalling conditions of thick mud, poor visibility and the ever-present worry of oxygen poisoning. The diver made several attempts to cut the cables but it could not be determined if the attempt was successful. It was not until after the war that it was confirmed that the cable had been cut. *XE5* and *Selene* safely returned to Subic Bay on 6 August 1945.

Adam Bergius joined the Royal Navy in 1942 and after his initial training at HMS *Ganges* was drafted to HMS *Agamemnon*, an auxiliary minelayer from the 1st Minelaying Squadron based at the Kyle of Lochalsh. He was employed laying the 'Northern Barrage', a minefield in the Iceland–Greenland gap. In 1944 he underwent officer training at HMS *King Alfred* in Hove, East Sussex, and emerged a 19-year-old midshipman.

Looking for a bit of excitement, in 1944 he volunteered for 'Special and Hazardous Service'. It is worth noting that the young men who volunteered for this particular line of work did not actually know what they were volunteering for, other than it was special and hazardous. It could have been the Long-Range Desert Group, being parachuted into France or any other perilous task the War Office deemed necessary. Eventually, Adam found himself on the way to Rothesay to join HMS *Varbel*, the home of the fledgling 12th Submarine Flotilla. Shortly after joining, Adam learned he was to train as a diver for the secret X craft. He had to practise getting in and out of the submarine using the wet and dry chamber. He also practised cutting underwater nets, placing explosives and many other tasks the diver might be expected to carry out. Most of this training took place at Ardtaraig House (*Varbel II*) on Loch Striven.

I first met Adam at the West of Scotland Submarine Association Branch's *K13* dinner in January 2013, where he was the after-dinner speaker. Below is the speech he gave detailing his experiences during the war:

> I am afraid it is a rather stone-age submariner you have to put up with tonight. You lads all came here by sea to Faslane but we came to Military Port No. 1, as it was known at that time, courtesy of London Midland and Scottish Railways, pulled by a steam locomotive with our submarine *XE4* heavily disguised as a food container on a wagon in tow and behind ourselves, CO Lt Max Shean from Australia, No. 1 Ben Kelly from Edinburgh, Chief ERA Ginger Coles from Newbury and myself, a diver from Argyll, in a first class carriage. It was November 1944 and the improved design XEs were rolling out of Barrow-in-Furness and 1–6 were to be ready to head for the Far East early in the New Year. It was a quick work-up at Port Bannatyne and then the new 14th Flotilla under the command of Captain WR 'Tiny' Fell was off for the Far East in the depot ship Bonaventure.
>
> It was a very pleasable sea voyage and with dreams of plenty of targets in much less heavily defended harbours in the many islands occupied by the Japs, all in needing food! There was only one incident of alarm when we were approaching the Caribbean Islands when there was a cry from the starboard look out, 'Strong smell of dusky maiden with flowers in her hair.' He was about to be rebuked by the officer of the Watch when all our nostrils were assailed by the scent and got a 'Thank you' before the signal to 'stop' was sent to the engine-room.
>
> 'Listen for breakers' was the next order. Captain was already on the bridge followed by the Navigating Officer who was my boss as I had been made Navigators Yeoman and had taken the evening stars just some seven hours before, which had shown us some 190 miles to the NE of Tobago. Bonaventure could not do 27 knots! And then as suddenly as the dusky maiden had come, she was gone!
>
> Then through the Panama Canal and on to Pearl Harbor, where because of the ever-present secrecy we were not allowed ashore but soon we got the news that we were not wanted by the US Navy. Many thought it was that our form of warfare was 'Not quite cricket'. I think the reality was that there

was a general feeling that the Pacific War was the domain of the US not the UK and what was worse [was that] we had a weapon they did not have. This was only a feeling at the top but at the lower levels of command our cheerful and determined CO was received with enthusiasm and respect and he was able to get permission to proceed to Australia, and what a place it proved to be for a diver! We anchored at Whitsunday Island on the Barrier Reef to do our warm water work-up. Instead, we found ourselves in a wonderland of colour and light. Corals and sea plants and fish of every shape and colour and a world quite unknown as any pre-war divers would have been in cumbersome suits with air pumped down to them from a boat on the surface. We could swim free as a fish and being on pure oxygen made no noisy bubbles to destroy the peace of the scene. But alas as we all know you get nothing for nothing and two experienced divers just disappeared and a third one was very lucky not to go the same way. It was when I was on a routine exercise I came across an old ship's anchor cable going down the outside of the reef. I pulled myself down it and was amazed to find an old sailing ship. There were two men on the foredeck and I was just about to speak to them (early signs of oxygen 'poisoning') when a well-known voice rang out, 'You're alright, carry on normal breathing Sir'. A command from CPO Harry Fright, our diving chief, as I was stretched out on the floorboards of the diving boat. A command which I have been happy to carry out for the following sixty-seven years!

However, in spite of all this the real purpose of our presence in these waters seemed to be vanishing in spite of many trips by Capt Fell up to the battle area in the Philippines at Subic Bay and many meetings with US submariners, particularly Rear Admiral Fife of Western Pacific Command Submarines, to try to get work for us to do. Things got gloomier and gloomier and morale was falling to a low level. Finally, we heard there was to be a meeting in Sydney with Admiral Lockwood I/C US Submarines Pacific and Admiral Sir Bruce Fraser to decide the fate of the 14th Flotilla and Bonaventure. It had virtually been decided to scrap the X Craft and convert Bonaventure to cargo carrying when a Captain Claibourne USN, who seemed to speak with great authority, asked if there was anybody who could cut submarine cables in enemy waters, to which Tiny Fell said, 'We can!' and so we went happily back to the Queensland coast to find disused cables [to practise with]! All sorts of different grapnels were made and tried out to achieve the maximum penetration of the mud and the minimum use of battery power. 'Group up, half ahead' was normally needed. However, it all seemed possible and at the same time came instructions from Fife for two boats to attack the cruisers *Takao* and *Myōkō* in the Johore Straights to defend Singapore. The decision was made much to our disappointment that this job was to be done by *XE1* and *XE3*. We were to cut two cables at Saigon, one to Singapore and one to Hong Kong, and *XE5* was to cut the cable to Japan at the Hong Kong end. As we had two to cut it was decided that there should be two divers and Ken Briggs, also an Australian, joined us. Five was quite a squash in an X boat! However, everyone was cheerful again and Bonaventure took us to Subic Bay to drop off *XE5* and we proceeded to Labuan Borneo flying the flag of Rear Admiral Fife and there to join up with Spearhead to tow *XE4* with her passage crew to Saigon, where we took over. After two or three runs in about 30ft of water there was an almighty crash. 'Ah,' said Max, 'that must be the wreck, now we know exactly where we are!'

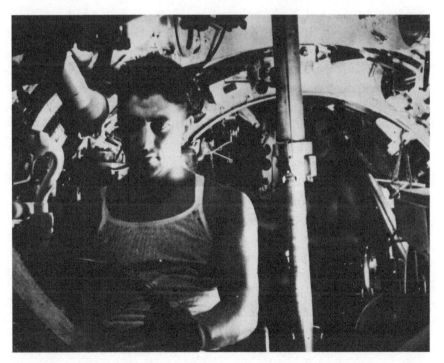

An interior view of an XE craft, with the periscope behind the helmsman. This gives some idea of the cramped conditions on board. (*National Museum of the Royal Navy*)

Shortly after we grabbed the first cable, then Ken was out to return very quickly with a foot-long sample of cable. *Saigon to Singapore away!* It was getting deeper all the time but the hope was that with the bottom getting harder the cable would be near enough to the surface to be caught, and then we got it at 50ft. Then it was out quietly, no excitement and there it was. In with the cutter and the damned thing jammed! Quietly back in again to take deep breaths of second-hand air to dilute the intoxicating O_2! Then out again with the second cutter and it was all done in jig time and back with the foot of evidence. *Saigon to Hong Kong away!* We saw no naval ships but the waters where we were was [sic] full of fishing junks. We were lucky not to be caught! By now the air was not good. I noticed it because I had been on the 'hard stuff' and Max ordered 'Up induction mast, periscope depth' and then 'Open induction, stop main motor, in engine clutch, half ahead'. Early days of the Snorkel but in the able hands of Ben's light touch he could keep the boat at a depth to the nearest six inches. Soon everyone was revived and in the darkness of the infra-red light from Spearhead. The plan had been to stay overnight and attach our limpets to shipping in Saigon harbour the following day but U.S. aerial 'recce' had revealed there was nothing worth attacking so it was back to Bonaventure again!

Shortly after this XEs 1 and 3 came in, Takao had been badly damaged but Myoko could not be reached. A further attack was planned to be carried out by XEs 3 and 4 so it was all go. We actually were in tow when the signal came through that we were to return and that an atomic bomb had been dropped on Japan!

None of us knew, what it was! The following day came the famous signal 'Cease hostilities against Japan'. Huge celebration and then off to Subic Bay to collect Percy Westmacott and *XE5*. They had had real problems with finding the cable in thick mud and were not sure whether they had managed to cut it or not. All very frustrating for them. Then it was under way for Sydney. I did not want to go and wanted to leave X craft and if possible stay in the interesting part of the World we were in. When we were at Manus, Admiralty Islands just north of Australia, Tiny Fell summoned me to tell me I had been appointed pilot of the steam tug Empire Sam to take a little convoy of sewage barges and small coastal craft to Hong Kong. It was sad to leave Bonaventure and the next time I was to see Tiny was on the ski slopes of the Cairngorms in 1949! But what a joy to be off on a 3,700-mile pleasure cruise through the Philippines.

When we reached Hong Kong the last thing a harbour tug needed was a full-time navigator and then to my joy I discovered three 'S' boats lying alongside: *Sidon, Spearhead* and *Scotsman*. Then I discovered that poor Sammy Large, the pilot of Spearhead, had been shot in the shoulder by a Jap soon after their arrival there and was still in hospital. So I got the job and a pretty leisurely one it was as we spent two weeks providing power for the Hong Kong tram service and one week at sea looking for Japs on the surrounding islands and also keeping an eye on the smuggling trade which seemed to be in good heart! Versatile weapons these submarines! Then more big celebrations and I was quite puzzled by the award of DSC for the small task I had carried out so recently.

It was to be some fifteen years later when I was in New York, an American friend asked me to a dinner, the Chairman being Admiral Kincade, who at the time of the cable cutting was I think C-in-C US 7th Fleet Pacific, and I have to say I was very embarrassed by the praise he lavished on *XE4* and her crew. Another few years were to pass before there was a programme on the BBC about the surrender in early May 1945 of a German freighter U-boat in the Atlantic, which was taken by Admiral Kincade. Her cargo comprised a lot of nuclear material, a complete broken down latest model Messerschmitt and a party of top Japanese scientists and Engineers, all of whom had committed suicide. It all must have given the feeling that Japan was in no mood to surrender and that Germany was still actively sharing her knowledge with her. It was just over two weeks later at the meeting in Sydney when the mysterious and obviously powerful Capt Claiborne came out of the blue to ask if anyone could cut the cables! A look at his past revealed that he was a fluent Japanese speaker, had made a deep study of Japanese attitudes and psychology and had served in the pre-war years as Naval Attaché to the US Embassy in Tokyo. Could it be that by 1945 he was on an advisory council to President Truman, who was faced with the dauntless decision whether to drop the bomb or be faced with endless loss of life on both sides from a full-scale invasion of Japan? At this stage US intelligence had cracked the codes used by Japan in their radio communications but all their signals to their huge forces on mainland Asia from Hong Kong to Singapore could go undetected by cable. If this intelligence became available to the de-crypters then the President would have the full picture before making his fateful decision. The bombs were dropped and all very sad for the large number of civilians killed but it demonstrated to the world just

This is the piece of the Saigon to Hong Kong telephone cable that was cut by Sub-Lieutenant Adam Bergius in Operation Sabre. (*Author's collection*)

what a brutal weapon now existed and allowed it to take its place as a silent and all-powerful deterrent to play its leading part in keeping the peace.

And that is what you lads do along with a lot of other things now. We have all done our bit in our time and have finished our time with a wealth of friends and memories. Ginger and I are the only survivors of *XE4* but we still keep in touch and are grateful for the friendship of a lot of other submariners and so we carry out the wishes of Capt. W.R. Tiny Fell, Capt. 12th Flotilla:

> To serve in submarines is to become a member of the strongest, most loyal union of men that exists. Scores of people ask, 'why did men join submarines and how could they stick in them?' There are many answers to that question. For adventure and fun at the outset; then because of the intense interest, and because of the variety of tasks that

must be at one's fingertips. There is every reason why he should join and delight in joining submarines, but the greatest joy of all is the companionship, unity and feeling that he is one of a team.

The two X craft bases were decommissioned in 1945: *Varbel II* on 28 February and HMS *Varbel* on 15 May.

The midget submarines and human torpedoes of the 12th Flotilla achieved an impressive wartime record. The flotilla was in commission for only three years, but during that time its crews were awarded four VCs, three CBEs, eleven DSOs, one OBE, ten MBEs, seventeen DSCs, six CGMs, eleven DSMs, four BEMs, and approximately 100 mentions in despatches. They sank 100,000 tons of enemy shipping; the disabling of *Tirpitz* enabled British battleships to be freed from Home Fleet duties and undoubtedly saved countless lives. The attack on the floating dock at Bergen restricted U-boat repairs and reduced the amount of U-boat activity off the coast of Norway. In the Far East, the attack on *Takao* and the cutting of the telephone cables undoubtedly saved countless lives and severely disrupted the Japanese war effort.

4

X CRAFT BUILT DURING THE SECOND WORLD WAR

X1   was intended to be a submersible commerce raider; at the time of its launch it was the largest submarine in the world. It was laid down on 2 November 1921 at HM Dockyard, Chatham, and commissioned in December 1925. It was broken up 12 December 1936.

X2   was the captured Italian submarine, *Galileo Galilei*. It was captured on 19 June 1940 in the Red Sea by British trawler *Moonstone* after submarine crew were overcome by toxic gases. It was renamed *P711* and broken up in 1946.

X3   unofficially named *Piker 1*. It was built by Varley Marine, Hamble. It sank on 4 November 1942 in Loch Striven due to a leaking engine valve. Fortunately, all the crew escaped by using their Davis Submerged Escape Apparatus. It was scrapped in 1945.

X4   built by Portsmouth Dockyard and scrapped in 1945.

Based on these two experimental midget submarines, operational craft were quickly developed and the next six X craft were built by Vickers-Armstrongs, Barrow-in-Furness, starting in December 1942. All were lost in the attack on *Tirpitz*, code-named Operation Source. The submarines' unofficial names are shown in brackets.

X5 (*Platypus*) sunk in the Altenfjord, 22 September 1943.

X6 (*Piker II*) scuttled in the Altenfjord, 22 September 1943.

X7 (*Pdinichthys*) scuttled in the Altenfjord, 22 September 1943. The German Navy raised the stern section following the attack. The remains were finally raised in 1974. Salvaged in 1976 for museum restoration at the Imperial War Museum, Duxford.

X8 (*Expectant*) scuttled in the North Sea, 17 September 1943.

X9 (*Pluto*) foundered under tow in the North Sea, 16 September 1943.

X10 (*Excalibur*) scuttled in the North Sea, 3 October 1943.

X11–X19 cancelled.

The next six X craft were built by small northern engineering companies, which were more used to producing tractors and other farming equipment.

X20 (*Exemplar*) built by Broadbent, Huddersfield, and used in Operation Postage Able (surveying the Normandy beaches prior to invasion) and on Operation Gambit (landing beach markers for the Normandy landings).

X21 (*Exultant*) built By Broadbent.

X22 (*Exploit*) built by Markham & Co., Chesterfield. Collided with HMS *Syrtis* and lost with all hands while training, 7 February 1944.

X23 (*Xiphias*) built by Markham and used on Operation Gambit. Sold in 1945.

X24 (*Expeditions*) built by Marshall, Gainsborough. Used on Operation Guidance, the attack on the floating dock at Bergen on 15 April 1944, when the merchant ship *Barenfels*, which was alongside the dock, was sunk. The dock was attacked a second time in Operation Heckle on 11 September 1944, again by *X24*, and sunk. The X craft is now at the Royal Navy Submarine Museum at Gosport.

X25 (*Xema*), built by Marshall, sold 1945.

During 1943 and 1944, Vickers built six XT craft for training purposes. These had an endurance of only 500 miles at 4 knots and were a simplified version of the X5 class. They were not fitted with the side cargo-release gear, night periscope or automatic helmsman. Also, their day periscope, projector compass and air-induction trunk were fixed in the 'raised' position. In May 1946 two – believed to have been *XT1* and *XT2* – were towed to Aberlady Bay, East Lothian, Scotland and used for trials with de Havilland Mosquito aircraft. These aircraft were used to train RAF crew for Coastal Command anti-shipping strike squadrons and they targeted the two XT craft to test the effectiveness of cannon shells in piercing their armour plate. Their remains can still be seen at low tide.

XT1 (*Extant*) built by Vickers and scrapped 1945.

XT2 (*Sandra*) built by Vickers and scrapped 1945.

XT3 (*Herald*) built by Vickers and scrapped 1945.

XT4 (*Excelsior*) built by Vickers and scrapped 1945.

XT5 (*Extended*) built by Vickers and scrapped 1945.

XT6 (*Xantho*) built by Vickers and scrapped 1945.

XT7–XT19 cancelled.

The XE-class submarines were a series of twelve midget submarines that were an improved version of the original X class. They carried a crew of four, typically a lieutenant in command, with a sub-lieutenant as deputy, an engine room artificer in charge of the mechanical side and a seaman or leading seaman. At least one of the crew was a qualified diver. In addition to the two side charges (each of which contained 2 tons of amatol explosive), they carried around six 20lb (9kg) limpet mines, which were attached to the target by the diver.

FIRST GROUP

XE1 (*Executioner*) built by Thomas Broadbent and Sons. Used in Operation Struggle, the British destruction of the Japanese heavy cruiser *Takao* in Singapore harbour. Scrapped in 1945.

XE2 (*Xerxes*) built by Thomas Broadbent and Sons and scrapped in 1945.

XE3 (*Sigyn*) built by Thomas Broadbent and Sons and used in Operation Struggle. Scrapped in 1945.

XE4 (*Exciter*) built by Thomas Broadbent and Sons. Used in Operation Sabre, the cutting of the Hong Kong to Saigon telephone cable. Scrapped in 1945.

XE5 (*Perseus*) built by Thomas Broadbent and Sons. Used in Operation Foil, the cutting of the Hong Kong to Singapore telephone cable. Scrapped in 1945.

XE6 (*Excalibur II*) built by Thomas Broadbent and Sons and scrapped in 1945.

XE7 (*Exuberant*) built by Thomas Broadbent and Sons and scrapped in 1952.

XE8 (*Expunger*) built by Broadbent. Sunk as a target 1952. Recovered in 1973 and preserved at Chatham Historic Dockyard, on loan from the Imperial War Museum.

XE9   (*Unexpected*) built by Markham and scrapped in 1952. Sunk as a target.

XE10 built by Markham, cancelled incomplete in 1945.

SECOND GROUP

XE11 (*Lucifer*) built by Marshall. Collided with a boom defence vessel in Loch Striven after drifting out of its exercise area and lost on 6 March 1945. Three crew were killed in the accident but two managed to escape. The boat was later salvaged.

XE12 (*Excitable*) built by Marshall and cannibalised for spares in 1952.

XE12–XE25 cancelled.

PART TWO

THE ORIGIN OF THE STICKLEBACK CLASS

5

THE COLD WAR

The British developed the Stickleback class of midget submarines to do one specific task during the initial stages of the Cold War: to combat a perceived threat posed by a former ally.

It is generally accepted that the origins of the Cold War can be traced to the period immediately following the Second World War. However, there is an arguable case to be made that its roots stretch back to the beginning of the Soviet Union, the October Revolution of 1917 when the Bolsheviks took power.

I suggest its roots go even further back than the October Revolution. Both America and Russia were born from revolution but developed differing attitudes in their dealings with the rest of the world due to their different and opposing ideologies. The American state had little influence over the day-to-day life of citizens and the Constitution limited its power. On the other hand, mainly due to the influence of Eastern Orthodoxy and the rule of the Tsar, Russia was a bureaucratic, land-based power that saw its security in terms of the land it owned. Furthermore, there were significant differences in the two countries' attitudes concerning empire building. The Americans and many other Western nations were primarily seafaring nations with trade-based economies. At the same time, Russia tended towards isolationism, viewing anything outside its direct control with suspicion. This was seen as a threat in the West. The differences between the two political and economic systems were simplified and eventually matured into national ideologies. Socialism versus capitalism, isolationism versus trade, and state planning versus private enterprise, became the main areas that differentiated and separated their two ways of life. The atheistic nature of Russian communism also concerned many Americans. This led to a fundamental cultural difference between Russia and America and East and West. In turn, these differences led to suspicion and have underpinned and undoubtedly adversely influenced the dealings between the two blocs over the years and provided a distorted prism that became the lens through which all events were considered and evaluated.

Most importantly, these culture-based differences manifest themselves in people's behaviour and attitudes. This is the primary cause of the troubles between East and West, as the war oscillated through its various phases. In Russia, the state comes first; in Western countries, it is the individual.

Regardless of its start date, the Cold War was a peculiar conflict. The two main participants never directly fought one another but this proxy war, in one way or another, affected the lives of most people in the world. It was a war without a clear-cut start, and despite the celebrations in 1991, it is a war that is still in progress.

The long and tangled history between Russia and the West started in 1553 when Richard Chancellor, the English navigator, landed at Arkhangelsk. He revisited Russia in 1595, when the Muscovy Company, which had the monopoly of trade between the two countries, was formed. The relationship between the two countries was further strengthened when the Russian Tsar, Peter I, visited England in 1697. He would invite English engineers to St Petersburg in the 1720s, where they eventually formed a small – but nevertheless important – 'expat' community.

During the eighteenth century, Britain and Russia were allies as often as they were on opposing sides in the various European wars. Britain's primary concern was the possibility of a Russian expansion into Ottoman Turkey; this would give the Russians a port in the Mediterranean and allow them to control access to the recently opened Suez Canal. This was to be a recurring theme in the following years. In addition, as the Russian Empire continued to expand, concern grew in Britain about the closeness of the Tsar's Empire to India. This led to 'The Great Game', a series of armed conflicts in Afghanistan, further increasing the tension between the two countries.

The Americans also had their own problems establishing a stable relationship with Russia. During the nineteenth century, there were several conflicts between them, all concerned with the opening of East Asia to American trade and influence. America's first national government appointed Francis Dana as Minister to St Petersburg in 1781. However, probably due to Russia's diplomatic ties with Britain, the position was never formally recognised. By the late 1700s, the Russians had become interested in America's Pacific Northwest, primarily for fur trapping and trading. By 1788, several Russian settlements had been set up in Alaska, including the mainland areas around Cook Inlet.

During 1799, the Russian America Company was established to develop trade opportunities in the North Pacific and in June 1810, the First Russian Minister to America was appointed. The Russians set up a settlement at Fort Ross on the west coast of North America, now Sonoma County in California, in 1812; at this time, this region was Spanish territory. It was the southernmost Russian settlement in North America from 1812 until 1842, when it was sold to America.

During the 1918–20 Russian Civil War, many countries, including Britain, France, Japan, Canada and America, supported the White Russians against the Bolsheviks. The revolution consisted of two separate coups, in February and October 1917, but it took a further three years before Lenin came to power. France and Japan invaded Russia in an attempt to remove the newly formed government. America landed troops in Siberia in 1918 to protect its interests; they also landed forces at Vladivostok and Arkhangelsk. Also, in 1918, Britain supplied money and troops to support the anti-Bolshevik 'White' counterrevolutionaries. This policy was implemented by the Minister of War, Winston Churchill, a well-known imperialist and anti-communist. Despite this, the Bolsheviks took complete control of Russia and its provinces, such as Ukraine, Georgia, Armenia and Azerbaijan. These actions undoubtedly coloured Russian attitudes in their dealings with the Western world over the coming years. After this period, tensions between Russia and the West became more ideological in nature.

Furthermore, Russia had borrowed heavily before 1917 to finance industrialisation and railways. At the time of the revolution, the country was the world's largest net debtor. The Bolsheviks blamed the Tsarist drive to industrialisation as failing the working class. Trotsky said of the Bolsheviks, 'They said "we are not paying and even if we could, we wouldn't pay"', essentially denying and cancelling the nation's debt. It was in the mid-1980s that Russia recognised some of this debt. The decision to renege on the debt further alienated Russia from the West, particularly France, which held a great deal of it and its banks and citizens suffered massive losses.

At the start of the First World War, America, Britain and Russia were allies. Then in 1918, Lenin negotiated a separate peace deal with the Central Powers (the Germans, Austro–Hungarian Ottoman Empires and the Kingdom of Bulgaria) at Brest-Litovsk. This left the Western Allies to fight the Central Powers alone, while allowing the Germans to deploy more troops to the Western Front. This episode further intensified Western mistrust of the Russians.

Western powers diplomatically isolated the Russian Government, causing Lenin to state that a 'hostile capitalist encirclement surrounded his country'. He thought of diplomacy as a weapon to keep Russia's enemies divided. To this end, he set up the Comintern, an organisation to promote sister revolutions worldwide by any means possible. Its primary objective was to spread the radical Russian message. However, it could have been more effective; it failed in attempts to start revolutions in Germany, Bavaria and Hungary. As a consequence of these failures, Russia became even more inward-looking and increasingly isolated from the international community. Also at this time, Russia was the most authoritarian society in the world.

In the West, this resulted in an intense anti-communist feeling, particularly in America; this became known as the 'First Red Scare' and it lasted from 1917 to 1920. During this period, there was a rising countrywide fear that a Bolshevik-type revolution in America was imminent and that this would have a devastating effect on the American way of life.

As I wrote in my previous book, *Polaris: The History of the UK's Submarine Force*:

> At the end of the First World War the British Government and the fledgling trade union movement disagreed on several issues, not least of which was what the relationship should be with Russia. While the Conservative government was refusing to have any contact with Russia, the Labour Prime Minister Ramsay MacDonald had signed several agreements with the country and he was in favour of establishing formal relations. MacDonald also proposed that the Russians should be given a large loan. Because of these actions there were claims from the opposition benches that the government had been infiltrated by many communist sympathisers and agents. It appeared that Parliament would reject MacDonald's proposals and as a result the Prime Minister called a general election. Any hopes of a Labour victory were crushed when a letter, allegedly written by Zinoviev (head of the Comintern) was sent to the British Communist Party. It seemed to suggest that MacDonald's proposals would be useful for both the British and Russian Communist Parties and, more damning still, it went on to suggest that the British workers, with the help of the military, should rise up against the government. In response, MacDonald replied, stating his disapproval, but despite this, he lost the election.
>
> …

Joseph Stalin became Russian leader after Lenin's death on 21 January 1924. He regarded his country as a socialist island and he felt he had to fight capitalism and spread his socialist message throughout the world. He saw international politics to be, essentially, a two-part process; the East and the West. Countries that were attracted to socialism would naturally be drawn to Russia, on the other hand, Western countries would be drawn towards capitalism. Needless to say, fear and mistrust grew between the West and Russia, and this was driven by several events, primarily the Bolsheviks' challenge to capitalism; in 1926 the Russians helped fund the British general workers' strike. A consequence of this was that Britain broke off relations withh Russia, and matters were not improved when in 1927 Stalin announced that peaceful coexistence with 'the capitalist countries is receding into the past'; Russia claimed French and British involvement in a coup d'état during the Shakhty show trial, while more than half a million Soviets were executed during the Great Purge. During the 'Moscow show trials' it was implied that Britain, France, Japan and Germany were involved in espionage against Russia. Other things that affected Russia's relationship with the West included the death of 6–8 million people in the Ukrainian Soviet Socialist Republic in the 1932–33 famine, Western support of the White Army in the Russian Civil War, America's refusal to recognise the Soviet Union until 1933 and the Soviet entry into the Treaty of Rapallo. The latter was an agreement Russia and Germany signed on 16 April 1922 that committed each country to renounce all territorial and financial claims against the other following the Treaty of Brest-Litovsk and the First World War. The countries also decided to normalise diplomatic relations and to 'co-operate in a spirit of mutual goodwill in meeting the economic needs of both countries' [Treaty of Rapello]. Just before the Second World War, as a result of Western appeasement of Adolf Hitler, the Russians arranged a series of meetings with the British and French in the hope of forming an alliance to counter the threat posed by Nazi Germany. This failed primarily due to British and French concerns regarding Bolshevism and socialist revolution.

The existing deep-rooted suspicion and mistrust of Russian intentions were further intensified when, on 23 August 1939, one week before the outbreak of the Second World War, Russia signed a non-aggression pact with Germany. This became known as the 'Molotov–Ribbentrop Pact', named after the two foreign secretaries involved. The pact included a secret agreement to split

Poland and Eastern Europe between the two states. The Allies saw this as a particularly underhand act designed to challenge their aims and seriously began considering military intervention. In addition, they increased their economic sanctions against the Bolsheviks.

On 1 September 1939, Poland was invaded by German and Russian troops. As a result of this, two days later, both Britain and France declared war on Germany.

Germany and Russia had a wide-ranging economic relationship, which included trading essential war supplies. The Germans were allowed to base a submarine depot ship in the Kola Peninsula and even used a Russian icebreaker, which allowed a German commerce raider access to the Pacific Ocean. Stalin also stated that the Russians would not see Germany defeated and would offer direct military aid if required.

Even before the war started, it was more than apparent that Britain would need American help, both militarily and in terms of material support, if it were going to stand any chance of defeating Germany. Churchill, himself half-American, spent a great deal of time and effort developing the relationship. Although it was essential for the war effort and undoubtedly helped Britain win the war, the relationship cost the country much of her wealth and, eventually, her empire.

Prior to America entering the war, President F.D. Roosevelt introduced a policy called 'Cash and Carry' on 21 September 1939. This permitted America to sell goods to warring nations, provided that these materials were carried in their own ships and the supplies were paid for in cash before they were collected. The Americans hoped that this policy would enable them to remain neutral while actually helping Britain. Germany had very little money and any goods they could manage to buy had to cross the Atlantic Ocean, which was controlled by the British Navy, which made transportation of the goods difficult, if not impossible. Meanwhile America believed that Germany might invade Britain, which would have made the British currency worthless.

To overcome this problem initially, British-owned companies were signed over to the Americans. Later, in Operation Fish, British gold reserves and securities were transported to Canada by the Royal Navy, then transferred to America. American Navy ships were used to transport British gold reserves from South Africa.

On 22 June 1941, Germany broke the Molotov–Ribbentrop Pact and invaded Russia. Russia and her soon-to-be Western Allies had little choice but to put aside their long history of mutual distrust and suspicion and work together to defeat their enemy, Adolf Hitler. In desperate times, the enemy of your enemy becomes your friend. The Americans were isolationists, the British were imperialists and the Russians were communists; put together, they were the most improbable political and military partners. A few weeks after the invasion, the Anglo–Soviet Agreement was signed on 12 July 1941. This was a formal military alliance in which both nations agreed to assist each other and not make a separate peace deal with Germany.

Not surprisingly, it had become clear, even before the end of the war, that collaboration, after the war, between the Allies would, at best, be limited. There had been, for example, disagreements on military tactics, especially the question of opening a second front against Germany in Western Europe.

Stalin wanted Britain to open a second front and invade northern France and raised this in July 1941. At this time, Britain just did not have the resources to do this, so refused the request. The Russians thought that the Allies had delayed the opening of the second front with Germany on purpose; they believed that the Allies intended to intervene at the last minute so they could influence the peace settlement and dominate Europe. They felt that by bearing the brunt of the assault, they were being left to weaken the German forces, which would make the eventual invasion of Europe – whenever it might happen – easier. This was an opinion they held onto throughout the Cold War. Whether or not the Russian rationale was correct or not, it only increased the already existing atmosphere of tension and hostility between East and West. On both sides it was only ever an alliance of necessity, born of need, not obligation or friendship.

Although the Americans shipped vast quantities of Lend-Lease material to the Russians, they did not join the European war until after the attack on Pearl Harbor on 7 December 1941. Britain signed a formal alliance with the Russians soon after the German invasion.

These goods were carried in convoys that initially sailed from Iceland to Arkhangelsk, or when winter ice closed the port, Murmansk. A total of 1,400 merchant ships sailed in seventy-eight convoys. They carried millions of tons of vital cargo and thousands of tanks, fighter aircraft, fuel, ammunition, raw materials and food. Unfortunately, eighty-five of these ships were sunk, causing the deaths of more than 3,000 Allied seamen. The Royal Navy lost one escort carrier, which was damaged severely, two cruisers, six destroyers and eight other escorts.

The German invasion of Russia was an exceptionally pitiless and vicious conflict, with both sides showing a level of cruelty that was almost primitive in its nature. Over 20 million Russians lost their lives during the war; thousands of their cities, towns and villages lay in ruins; and more than 30,000 factories were destroyed.

By the end of the war, the Russians occupied Eastern Europe while Western Europe was in the hands of the Allies. As agreed at the Yalta Conference, the Americans and the Russians set up zones in Germany, and the Russians also moved troops into Poland, Hungary and Czechoslovakia. During this period, Russia was attempting to form a 'puppet' state in the Caucasus (northern Iran), in spite of the assurances they had given their allies. Also, in order to gain access to the Bosporus for their Navy they put pressure on Turkey and moved into the Balkans while installing puppet governments in Romania and Bulgaria, as well as supporting communist parties in Yugoslavia, Albania and Greece.

Stalin intended to destroy Germany's industrial capabilities to prevent the country from remilitarising and he wanted Germany to pay massive and totally unrealistic amounts of money in war reparations. This concerned the Western Allies, who thought that a strong Germany was essential to the recovery of Europe. The Russian actions and attitude increased fears that they were trying to build an empire that would threaten the recovering European countries. There was also strong opposition to Russian control over the buffer states. The American President Harry S. Truman, however, wanted precisely the opposite. He believed that only industrialisation and democracy would ensure post-war stability in Germany and throughout the European continent. As the two superpowers were unable to compromise or find common ground, they were set on a collision course.

Churchill was worried that the Americans would revert to pre-war isolationism. At the Yalta Conference, President Roosevelt had stated that after Germany's defeat, all American troops would be withdrawn from Europe within two years, leaving an exhausted Europe unable to defend itself against Russia. There was a possibility that the alliance would totally break down and the allies might even go to war against one another.

The Russians' hostile stance caused Churchill, in early 1945, to task the British War Cabinet's Joint Planning Staff Committee with developing an appropriate response. They proposed a plan 'to impose upon Russia the will of the United States and Britain': Operation Unthinkable. Essentially, America and Britain would continue fighting but now the enemy would be Russia. The plan received little support. Given the last six years, no one wanted to continue fighting.

The belligerent Russian stance left the Americans with little choice other than to react. Their initial response was the Truman Doctrine, where the Americans would support countries that were trying to resist a communist takeover, be this from internal armed minorities or from outside pressure. The American Secretary of State, George Marshall, proposed that loans should be given to help reconstruction and economic recovery in the devastated European states. This would also help develop a potential market for American companies, as they were the only ones that could supply the goods that war-torn Europe required. The sights Marshall had seen during his visit to post-war Europe deeply affected him and unquestionably influenced his proposal. The Americans made $400 million available to European countries in the form of economic and military aid. The policy was certainly a success and in 1953 he was awarded the Nobel Peace Prize.

The Americans were very worried that some Western European countries, particularly France, Italy, Greece and Turkey, would turn to communism because of the grave economic conditions that prevailed in their countries. The aid package helped the Italian Christian Democrats win the 1948 elections, defeating the powerful Communist–Socialist Alliance, and encouraged France to turn away from the communist party. Truman also gave $400 million dollars to Greece and Turkey: the Greek military went on to win its civil war and Turkey agreed to host several American missile bases. The plan was so successful that within just a few years, factories in Western Europe were exceeding their pre-war production levels. The Marshall Plan was initially available to all European countries, even the ones under Russian control, but Stalin referred to the American aid as 'dollar imperialism' and forbid communist countries from taking part. The Russians proposed their own version, which was intended to promote trade within Eastern Europe; this became known as the 'Molotov Plan'.

It seemed unlikely that the conflict between the two superpowers could have even been avoided. During the war, there was never any real relationship; it was more of a union based on need and convenience. There was no shared culture, rapport, comradeship or true friendship. As the guns fell silent in 1945, the suspicion and mistrust that had defined the East–West relationship for many years reappeared and the different ideologies and visions of the post-war world ensured they could not work together. The next war had already started or, perhaps more correctly, entered a new phase: the Cold War. But now there was a new complication, a deadly problem that was made worse by long-established mistrust and animosity; now, there were nuclear weapons and egos.

After the war, Stalin thought that the West would resume its historic colony and trade-based rivalry and this would leave Russia to continue its expansionist ambitions. The Russian economist Eugen Varga suggested that American cuts to military spending would result in overproduction, which, in turn, would lead to another depression. Based on this advice, Stalin expected the Americans would offer his country aid to help in post-war recovery, assuming that the Americans would need markets to maintain the current levels of industrial production that had significantly increased during the war. As American financial investments in its industry were maintained, this avoided any post-war problems with overproduction.

Even though Stalin joined the Americans in founding the United Nations, he opposed them on nearly every other programme, as they worked towards rebuilding Europe and establishing a new international order.

He objected to the Marshall Plan and the founding of the World Bank and the International Monetary Fund (IMF). In retaliation, he implemented his plan to create a buffer zone between Russia and Germany and established communist-friendly governments in Poland and other Eastern European countries. As a result, the so-called 'Iron Curtain' soon divided East from West in Europe. Stalin also tried unsuccessfully to drive French, British and American occupation forces from the German city of Berlin by blocking road and railway access. Determined not to let the city fall, Truman ordered the Berlin Airlift to bring in food and medical supplies for starving Berliners. Between 26 June 1948 and 30 September 1950, over 2 million tons of coal, food and other supplies were delivered by American and British aircraft. This episode unquestionably set the tone of suspicion, distrust and cynicism that would come to govern the attitudes of the two superpowers in their dealings with one another over the coming decades.

Polaris takes up the story from here:

> During 1946, while serving in Moscow, the American politician George F. Kennan wrote what became known as the 'Long Telegram' and in this he outlined what he believed were the Russian objectives. He argued that the Russians were motivated by deep-rooted imperialism and Marxist ideology; this caused them to be expansionist and paranoid, and as such they

presented a threat to America and its Western allies. In July 1947, he further developed his views in an article 'The Sources of Soviet Conduct in Foreign Affairs'. So powerful were the arguments that Truman based his policy of containment on Keenan's interpretation of Russian intentions. Later in the year the Soviet Central Committee secretary Andrei Zhdanov declared that the Truman Doctrine was 'intended for accordance of the American help to all reactionary regimes that actively oppose to [sic] democratic people, bearing an undisguised aggressive character'.

Tensions over Germany escalated after Truman refused to give the Russians reparations from West Germany's industrial plants. He rightly believed that it would inhibit Germany's economic recovery and the Western Allies felt a strong Germany was essential for European economic revival. In retaliation Stalin made the Russian sector of Germany a communist state. It was not until 1951 that the dismantling of West German industry was finally brought to an end, when the country agreed to place its heavy industry under the control of the European Coal and Steel Community, which in 1952 took over the role of the International Authority for the Ruhr.

NATO was established on 17 March 1948, when Belgium, the Netherlands, Luxembourg, France and Britain signed the Treaty of Brussels, which founded the military alliance and looked to the Americans to help it to counter the Russian threat. This agreement was ratified in the North Atlantic Treaty of 1949. The five original countries were joined by the USA, Portugal, Italy, Denmark, Norway, Iceland and Canada over the next few years.

Nikita Khrushchev became the Russian leader in 1956, some three years after Stalin died. He was communist of the old school and did not want his country's superpower status to decline – and so, unfortunately, the hoped-for improvement in relationships between America and Russia did not materialise. The Hungarian Uprising and Khrushchev's rather worrying habit of threatening the West with total nuclear annihilation undoubtedly played a role in the deteriorating relationship between the two superpowers. However, regardless of his threats, Khrushchev, unlike Stalin, thought that war between America and Russia could be avoided. Similar to Lenin, he was of the opinion that communism and capitalism could live side by side in peaceful harmony. That aside, he still thought that the Western capitalist system's inherent self-indulgence and greed would lead to instability that would cause the system to fail without any assistance from his own country.

6

BRITAIN AFTER THE SECOND WORLD WAR

The war had effectively bankrupted Britain. The national debt rose from £760 million to £3,500 million. Britain had spent almost £7 billion, which was a quarter of the national wealth, on the war effort. Once the Lend-Lease programme finished, Prime Minister Clement Attlee had little option but to send British economist John Maynard Keynes to America to attempt to secure financial help. Much to his credit, despite his ill health, he managed to secure a loan of US$ 3.75 billion. The Canadians also loaned Britain C$1.25 billion. The country also received £2.4 billion from the European Recovery Program (the Marshall Plan).

This was the largest share of the aid money, much more than any other European nation; Britain received over a third more Marshall Aid than West Germany. But whereas Germany used its share of the money to rebuild its industrial base and modernise its infrastructure, the Labour Government had different priorities. Government papers show that the primary use of Marshall Aid, at this time, was to maintain the Bank of England's reserves of gold and dollars, so that the country could continue to be the banker to the Sterling Area. The aid was also used to maintain overseas garrisons and imports, particularly food and tobacco, and the vast amounts of timber the government needed for its council house building programme. Spending capital investment in industrial and infrastructure modernisation was referred to as clearly of great importance.

Arguably, a substantial portion of the available money was squandered on maintaining oversea garrisons and supporting the country's currency against the gold standard. The government spent £2,000 million a year abroad, earning only £350 million in return. A third of all houses had been destroyed by bombing, and large numbers of factories and shops had also been destroyed. Britain endured 264,433 military and 60,595 civilian deaths during the war. Many people were physically and mentally damaged by the war and unable to resume a normal life. In addition, 177 merchant ships and two thirds of the Navy had been sunk, so food supplies were still a problem. Income tax remained high to help the government pay for the reconstruction. This was truly an age of austerity, not at all a country that thousands of recently demobbed servicemen had fought for, certainly not a land fit for heroes.

Despite the loans, the government had to adopt other money-saving solutions. Attlee's government introduced import controls and continued the wartime practice of rationing food; in fact, it got worse between 1947 and 1948, when about half of consumer expenditure on food was spent on rationed items. Food items, such as meat, cheese, eggs, fats and sugar were rationed; bread was rationed in July 1946, and potatoes in November 1947. Rationing was slowly relaxed between late 1948 and 1954, but coal remained rationed until 1958. The government also implemented defence cuts, which caused Labour deputy leader Herbert Morrison to express his concern in November 1949, stating the country could be paying more than it could afford for an inadequate defence organisation. The government also reduced capital investment affecting roads, railways and industry. The government also embarked on a massive programme of nationalisation, affecting approximately 20 per cent of the economy. On top of this, it created the very expensive National Health Service and embarked on an equally expensive nuclear weapons programme. Added to this were the approximately five million British service personnel who were demobilised and needed to be reinstated into society and found jobs and homes.

During the war, with thousands of men away serving their country, British women took on the men's jobs. They also coped with their traditional role of running the family home, dealing with rationing and growing food for their families in allotments and gardens. Women were called up for war work, and from 1941, they worked as mechanics, engineers, munitions workers, air raid wardens, and bus and fire engine drivers. Initially, only single women aged 20 to 30 were called up, but by mid-1943, nearly all women, single or married, were employed in factories, on the land or in the armed forces.

Despite the loans and the government's best efforts, by 1947 the country's economy was in crisis. At the TUC conference in 1948, the Labour chancellor Sir Stafford Cripps announced, 'There is only a certain-sized cake. If a lot of people want a larger slice, they can only get it by taking it from others.' He had already introduced a wage freeze in his austerity budget the year before. The government's problems were further compounded when the winter that year turned out to be one of the worst on record and the wheat harvest failed, adding to the rationing problems and further increasing difficulties with the country's food supply.

In 1949 the government had no choice but to devalue the pound; from US$4.03 to US$2.80.

The country was a grey, dark place with a depressing atmosphere during this period. There were problems with transport, factories ran out of fuel, and householders had restricted electricity supplies to light their houses or with which to cook. And on top of this were government-sponsored adverts advising people what to do in case of a nuclear attack.

People were guarded and suspicious; spies seemed to be everywhere as the Russians tried to obtain highly classified information about Britain's nuclear weapons and nuclear submarine programmes. Of these, the 'Cambridge Five' were the most famous. Four of them – Donald Maclean, Guy Burgess, 'Kim' Philby and Anthony Blunt – were recruited by Arnold Deutsch, a serving KGB operative, as Russian spies while at Cambridge University in the 1930s. There may have been a fifth spy in the ring, possibly John Cairncross. During the war and early 1950s, the men passed confidential information to the Soviet Union. In 1962, the British Government jailed John Vassal. He worked in the British Embassy in Moscow, but he was blackmailed into becoming a Russian spy. In 1961, three men and two women were jailed for plotting to hand information about Britain's first nuclear submarine to the Russians; they became known as the 'Portland Spy Ring'. The same year, George Blake was sentenced to forty-two years in prison. Blake had worked for British Intelligence but was a Soviet double agent. He escaped in 1966. Klaus Fuchs, a German-born British theoretical physicist, and Alan Nunn, a British physicist, both of whom were involved in the Manhattan Project, passed information to the Russians.

The empire had come to Britain's aid during the war and now wanted its independence; the country began dismantling the British Empire when India and Pakistan were granted independence in 1947, followed by Burma (Myanmar) and Ceylon (Sri Lanka) in 1948.

Britain was now a very different country than it had been before the war.

7

THE STICKLEBACK CLASS INTRODUCTION

Against this sombre background, the country's armed services instigated a series of comprehensive and systematic reviews and studies to access the full implications of the advances in science and technology that had been made during the war. For example, what would their role be in the post-war world, would there be a peace dividend and could they use this new nuclear technology? How might this new technology affect their operations and strategic thinking? It's interesting to note that during the 1950s, one important characteristic of the country's defence policy was a firm commitment to be at the head of all areas of military technology; this included the development of nuclear weapons, guided missiles, defence electronics and aircraft construction.

The government also conducted similar reviews, which in 1948 resulted in the 'Three Pillars Strategy'. These three pillars were: the defence of the country itself, maintaining vital sea lanes open and securing the Middle East as a defensive and striking base against Russia. These had to be in place to guarantee the country's security. This policy was followed in 1950 by the Defence Policy, which underlined the deterrent value of the American nuclear superiority over the Russians and the requirement to increase NATO forces to a level capable of preventing a Russian invasion of Western Europe, including a formal commitment to the defence of Germany.

In 1952 a second policy was published, the Global Strategy Papers. These were internal reviews led by the service chiefs and gave much more emphasis to the concept of a 'hot war' as opposed to a 'cold war' and the balance of conventional forces was shifted towards land and air forces.

In the naval reviews, the Director of Naval Air Warfare and Training (DAWT) suggested that the Navy could deliver atomic weapons well beyond the range of missiles launched from Britain. The Director of Naval Operational Research (DNOR), very prophetically, thought it might be possible to launch nuclear-armed rockets from submarines. Interestingly, the general conclusion was that the armed services' primary aim in the future might be the prevention of war rather than fighting it.

In part, these studies were hampered by the lack of information, particularly regarding the new nuclear technology.

Although British scientists had been involved in the Manhattan Project (the American-led Second World War project that led to the first nuclear weapons), there was very little information about this new technology, particularly about the atomic bomb itself, and what was known was confined to a small number of people. The first real information became available in August 1945, when the Smyth Report was published in America; it contained a basic account of nuclear physics and information about the bomb project itself. Several reports from British observers in Nagasaki and Hiroshima quickly followed this. Over a three-month period, the Americans allowed several British officials to collect information from the bomb sites. Their report, which mainly concentrated on the effects on buildings, was of little interest to the military reader, who was more concerned with the impact on shipping, military and air installations. The report had an annex detailing the effects of the explosions on light shipping in rivers and harbours.

The American nuclear tests at Bikini Atoll between 1946 and 1958 supplied subsequent information. Many foreign observers were invited to witness the tests, several even from the USSR, although the British seem to have favoured access and six of the eight members of the British delegation were from the Admiralty. This was followed by the Able and Baker tests, which resulted in a report that showed the expected results of an air burst and underwater explosion in various scenarios, including a naval dockyard (Portsmouth), a busy civilian port (Liverpool) and a fleet at sea. This showed that dockyards were particularly susceptible.

As noted in *Polaris*:

> It was the promise of high speeds and unlimited endurance that first kindled Flag Officer Submarines (FOSM) interest in nuclear power.
>
> To this end, in 1946, the Admiralty appointed several naval officers and member of the Royal Naval Scientific Service to the Atomic Energy Research Establishment at Harwell. A little later, during 1950, the Defence

Research Policy Committee (DRPC), along with the heavy electrical manufacturing company Metropolitan–Vickers, carried out a study into the potential of submarine nuclear power. The Treasury approved £500,000 for this research although the study was eventually suspended when it became apparent that the available nuclear fuel would be required for the nuclear weapons programme. However, in 1954 the Atomic Energy Authority stated that the necessary fuel could be made available. As a result of this, the Naval Section at Harwell (Naval Cell), under Captain Harrison-Smith, was placed on a more formal footing and work began in earnest. During the year the staff were increased by personnel from the Royal Corps of Naval Constructors (RCNC) and engineers from Vickers–Armstrongs Limited. Despite this, British interest was primarily centred round the German-inspired Hydrogen Peroxide (HPT) Steam Turbine, which eventually went to sea in two Ex-class submarines, HMS *Excalibur* and HMS *Explorer*. At this time the Russians were about to deploy their own nuclear submarines. It was intended that the high-speed peroxide submarines would enable anti-submarine forces to gain experience in tracking and attaching high-speed underwater targets.

Britain exploded its nuclear bomb, Hurricane, on 3 October 1952 in a lagoon off the western shore of Trimouille Island, Western Australia. This led to the Blue Danube plutonium bomb, Britain's first deployed weapon, in November 1953. It was intended to have 200 nuclear weapons by 1957, so plutonium production was increased by adding two new dual-use (plutonium and electricity) Magnox reactors at Calder Hall in Cumbria. With the delivery of the first Blue Danube free-fall atomic bombs to RAF Wittering in 1953, the RAF became the guardian of the British nuclear deterrent. For the next sixteen years, it held this role, one that was a defining feature of RAF operations through the late 1950s and '60s until June 1969, when the Royal Navy's Polaris submarines adopted the role. This was partly due to improvements in the Russian air defence systems, making it clear that a ballistic missile was the only way the country could maintain a credible deterrent.

Not to be left behind, the Army had its own tactical nuclear weapon project in the 1950s. Named Blue Peacock, renamed from Blue Bunny and originally Brown Bunny, the project's objective was to store several 20-kiloton nuclear land mines in Germany that would be placed on the North German Plain. Then, in the event of a Russian invasion from the East, they would be detonated by wire or an eight-day timer and would deny access to the area to the enemy for an long period due to radioactive contamination, as well as destroy facilities and installations over a large area.

In July 1952, the Director of Plans, J.J. Hopkins, submitted a paper to the Naval Board. Its purpose was to obtain approval for the long-term policy for the development and wartime employment of X craft.

The report suggested three roles for these in wartime:

A. To test and exercise the country's harbour defences and expand the number of X craft crews defences of British and Allied ports.
B. To attack enemy submarines, cruisers, floating docks and other important targets in enemy-held ports.
C. To land agents in enemy territory and carry out beach reconnaissance.

There were two long-term aims. Firstly, to develop X craft designed to keep ahead of the enemies' defensive techniques. Secondly, maintain, in peacetime, sufficient craft and trained crews to expand rapidly on the outbreak of war to the strength that towing facilities (available large submarines) and the targets would justify.

The report concluded that:

Four more X craft should be laid down after those planned for 1953–54 and twelve maintained in peacetime.

On the outbreak of war, a building programme of six X craft every three months should be started and continued until experience points to a revision.

The first priority for X craft in war should be to train Harbour Defences of this country and allied countries and expand X craft crews.

Offensive operations should not be started until we have sufficient craft to achieve really worthwhile success against U-boats unless a particularly attractive target is presented by the enemy.

When offensive operations start, they should take the form of frequent attacks by a single craft without preliminary reconnaissance unless a known valuable target justifies a 'set piece' attack.

The last paragraph of the report stated:

This paper has not considered the possibility of atomic explosives being available for the delivery by X craft. Should such explosives be available on the outbreak of war, it is believed that no delay in their use would be justified. On the other hand, should they be expected to be available within reasonable time after the outbreak of war, no previous attacks liable to alert the enemy's defences would be justified.

In almost a flashback to the primary role of their Second World War predecessors, documents newly released by the Public Record Office (PRO) show that the Royal Navy planned to build midget submarines capable of planting nuclear weapons inside Russian harbours. The proposed submarines would carry their 'cargoes' to the Kola Peninsula, the home of the formidable Northern Fleet, and once dropped, a delayed timer would detonate them.

This proposal, which was code-named Operation Cudgel, would address a major concern of the post-war Navy. The only method of delivering the nuclear bomb was by air via the RAF's V-bomber force and with this responsibility came the associated funds, resources and prestige. The Admiralty viewed this with more than a little unease. The historic rivalry between the Royal Navy and RAF was at a post-war peak. With the Americans being years away from being able to launch a nuclear missile from under the sea, Cudgel offered the Navy a way of getting a share of the huge budget that Britain was devoting to nuclear strategy at the start of the Cold War. Also, Britain's first nuclear test, in 1952, was conducted in a natural harbour in the Monte Bello Islands off Australia and provided first-hand information on the effects of placing a bomb in this type of target. The 'atomic' X craft design was first discussed in 1955; at this time, the Navy was still at least ten years away from launching a submarine or surface vessel that could launch a nuclear weapon.

The PRO files also highlighted the high degree of secrecy surrounding Cudgel. Several notes in the pack that was circulated around the various Admiralty departments concerned with the project were written by hand because there were not enough typists with the necessary top security clearance. The Director of Undersurface Warfare, Captain P.J. Cowell, told colleagues that: 'the existing X craft as built today is too complicated and too large for the purpose. A specially designed craft is necessary, whose sole function is to deliver the atom bomb and return the crew to the parent [submarine].'

The PRO documents detail the requirements for the X craft; they show that the vessel would be towed to a drop-off point off the enemy coast by a

P- or O-class submarine. Then the submarine and its two-man crew would travel up to 150 miles to its target. The crew would then prime, arm and detach their 'Red Beard' nuclear weapon. According to the documents, the weapon would float in relatively shallow water and be timed to detonate up to a week after being delivered. It was specified that the submarine should be able to operate in very low temperatures, showing that the primary targets would be the Arctic Ocean ports of Russia, for example, Murmansk and Arkhangelsk.

These X craft would have been ready by 1959, giving the Royal Navy a nuclear capability some six years before it actually achieved it. The project seems to have got off to a good start, with the Atomic Weapons Research Establishment at Harwell reporting that there was no reason why an atomic bomb should not be adapted for the purpose. But there was bad news for the proponents of Cudgel when A.J. Sims, the director of naval construction, wrote that a 30-ton craft, the most that could be carried by a conventional submarine, built using a non-magnetic material and with the endurance required, could not be made.

Despite this, further discussions took place into 1956 and detailed drawings of the new craft seem to have been made. But then an abrupt halt was called when in July of that year, Patrick Nairne, Secretary to the Board of the Admiralty, wrote to Captain Cowell to report a meeting of the Navy's masters:

> I am to acquaint you that Their Lordships have approved the following policy with regard to X craft.
>
> (a) Development of neither nuclear nor conventional weapons for X craft is justified.
> (b) Present X craft are to be retained and continue to operate.
> (c) No further X craft will be built.

PART THREE

BUILDING THE STICKLEBACK CLASS

8

CONSTRUCTING THE STICKLEBACK CLASS

The Royal Navy ordered an improved version of the wartime XE-class submarines. This new class of midget submarines was known as the Stickleback and their primary aim was to test British defences against Russian midget submarines. They were all constructed at Vickers Shipbuilding and Engineering Ltd (VSEL), based at Barrow-in-Furness, Cumbria.

The company was originally founded in 1871 by James Ramsden and was known as the Iron Shipbuilding Company but quickly changed its name to the Barrow Shipbuilding Company. In 1897, the company was bought by Vickers & Sons, which acquired the Maxim Nordenfelt Guns and Ammunition Company and became Vickers, Sons and Maxim Limited. The Barrow shipyard became the Naval Construction & Armaments Company. The company was renamed Vickers Ltd in 1911. In 1927 the company merged with Armstrong Whitworth and became Vickers-Armstrongs Ltd. The Armstrong Whitworth shipyard at High Walker on the River Tyne became the 'Naval Yard'. The shipbuilding division changed to Vickers-Armstrongs Shipbuilders Ltd in 1955, and changed again in 1968 to Vickers Limited Shipbuilding Group.

The four submarines were improved versions of the older XE-class submarines. They were primarily designed to allow British defences to practise defending against midget submarines, since it was thought that the Russians had or were developing this type of craft:

X51, HMS *Stickleback*. Yard no. 1037. Launched on 1 October 1954.

X52, HMS *Shrimp*. Yard no. 1037. Launched in October 1954 and scrapped in 1965.

X53, HMS *Sprat*. Yard no 1037. Launched on 30 December 1954 and scrapped in 1966.

X54, HMS *Minnow*. Yard no 1037. Launched on 5 May 1955 and scrapped in 1966.

A hull section under construction. (*The Dock Museum, Barrow-in-Furness*)

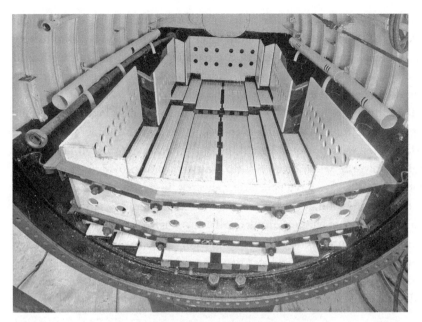

A battery compartment. (*The Dock Museum, Barrow-in-Furness*)

A battery compartment with batteries in place. (*The Dock Museum, Barrow-in-Furness*)

Similar to the earlier classes, the boats were of a relatively simple design. There were just four separate compartments: an engine room, control room, battery compartment and the wet and dry compartment, from which a diver could leave or enter the submarine to attach limpet mines to a target or cut nets. This also provided the main access to the submarine.

The submarines had an overall length of 53ft 10in with a beam of 6ft. They displaced 35.2 tons on the surface and 39.27 tons when submerged. On the surface, they were powered by a single 50hp Perkins P6 (6-cylinder diesel) engine, however, when dived a 44hp electric motor provided power to the propeller. It also acted as a generator, powered by the diesel engine to charge the battery. The submarines had a surface speed of 7 knots, a submerged speed of 6 knots, and a maximum diving depth of 100m. They had a surface range of 860nm and a submerged range of 80nm. They were crewed by five personnel; a captain, navigator, engineer and two divers. Unlike their larger sisters, they carried two 2-ton time-fused amatol charges (side cargoes) that were released under the target, and limpet mines that the divers placed.

A centre section, control room and wet and dry compartment. (*The Dock Museum, Barrow-in-Furness*)

An engine room section. (*The Dock Museum, Barrow-in-Furness*)

They were built in three sections: the engine room and battery compartment were separated from the control room and the wet and dry compartment. The sections had a joint ring flange at their open ends with approximately 160 bolts to join the units together. This design had the significant benefit of making it considerably easier to fit equipment into the hull, as access was possible through the open ends of the hull sections.

The frames for the three sections were assembled, preformed steel plates were welded to each other and then the submarine's frames.

Looking aft from the wet and dry compartment. The helmsman's seat is on the left in the foreground and the periscope can be seen behind him. This gives some idea of the cramped conditions on board. (*The Dock Museum, Barrow-in-Furness*)

The planesman's position. The trim and ballast control level and hydrophone can be seen in the foreground. The hatch to the engine room can be seen at the rear of the compartment. (*The Dock Museum, Barrow-in-Furness*)

The control room looking forward, with the periscope in the foreground. The hatch to the wet and dry compartment can be seen just to the left of the helmsman seat. (*The Dock Museum, Barrow-in-Furness*)

Then the larger pieces of equipment were put in place, followed by the various tanks. The pipes were fitted in the good old submarine tradition; the first fitter put his section of pipe in straight and then everybody else had to bend their pipes around the first one. The electrical wiring was installed and the various gauges, dials and many valves that were required to operate the submarine. Once this was completed, the seals were inserted between the three sections and then they were bolted together. Finally, the various systems that ran through them were connected.

The completed submarines were then moved by train or truck to their new operational bases.

9

LAYOUT

Submariners have always said that when a submarine is built, all the equipment is put in first and then the crew are squeezed in around it. Unfortunately, the X craft took this to a new level and this latest class did nothing to dispel this long-held belief. The vessels were incredibly cramped, with all available space filled with equipment or the few stores the submarine could carry. Added to this was the fact that it was impossible to stand upright anywhere on board. The crew were the last consideration.

Walking through the submarine, starting aft:

ENGINE ROOM

The engine room housed the single 50hp Perkins P6 (6-cylinder diesel) engine, which had an air start. This powered the submarine on the surface and when dived a 44hp electric motor provided power. On the surface, the electric motor acted as a generator to charge the battery and was powered by the diesel engine. There was a clutch between the diesel and electric motor, so the diesel engine could be disengaged when the submarine was in battery drive.

The submarines were fitted with a short 'snort' mast for surface running; this supplied air to the diesel. This was needed as the wet and dry compartment hatch was usually shut when the submarine was running on the surface. The diesel could not be run when the submarine was dived.

The submarine's hydraulic plant and its associated air bottles were sited in the engine room, as were the hydraulic pistons that controlled the rudder and hydroplanes.

The compressor was in the engine room and was used to recharge the submarine's compressed air bottles. These supplied the high-pressure air that was used to blow the water out of the ballast tanks, enabling the submarine to surface.

The aft trim tank, freshwater tanks and fuel tanks were positioned in this compartment.

The warmth from the engine made the engine room the only place on board where clothing could be dried. The crew wore heavy sweaters, vests and long johns under their Navy battledress or diving suits. They had to be careful that the drying garments did not block the air intakes, which could stop the engine.

There was a Brown A Gyro compass, which was in the engine room. Before the First World War, gyroscopic compasses were imported from Germany; British engineer Sidney George Brown started his company, S.G. Brown Ltd, in 1903.

CONTROL ROOM

This was the largest compartment on board and contained all the equipment required to operate and control the submarine. From here, the vessel was steered and the hydroplanes controlled its depth. There were valves to control the flooding and pumping of various tanks and a variety of dials and gauges that indicated tank contents. There was also a depth gauge, a radio and a hydrophone, which allowed the submarine to 'listen' to its underwater environment. There was a bunk in the control room that the captain used, it also doubled as a chart table.

In the centre of the control room was the periscope, which allowed the captain to navigate or confirm the position of the target when the submarine was dived. The periscope was raised and lowered by wires and an electric motor.

There were two seats in the control room. The hydroplanes operator used the aft one; the trim and bilge pumps were also controlled from this position. The trim control was like a car gear selection lever with several slots. Pushing the lever away discharged water from the submarine to the sea. If the operator pulled the lever towards him, water would be flooded into the tanks. Pushing the lever right or left would pump water between the aft trim tank and for'd trim tank. The engine control was next to this. The rudder could also be controlled from this position when required. The second seat was by the forward bulkhead, just to the right-hand side of the hatch into the wet and dry compartment; from here, the rudder was controlled. There was also an auto helmsman, which automatically kept the submarine on course.

LAYOUT

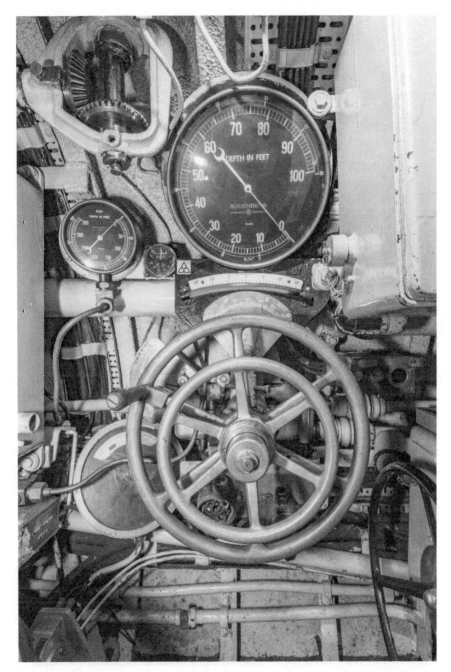

The planesman's position. The rudder control wheel can be seen on the left. *(Submarine Centre, Helensburgh)*

A battery compartment. *(Submarine Centre, Helensburgh)*

The planesman's seat. The hatch to the engine room can be seen in the background. *(Submarine Centre, Helensburgh)*

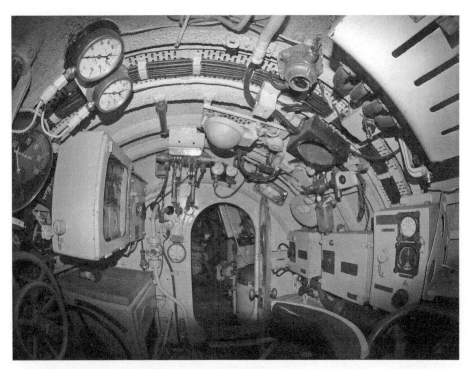

Aft in the control room. *(Submarine Centre, Helensburgh)*

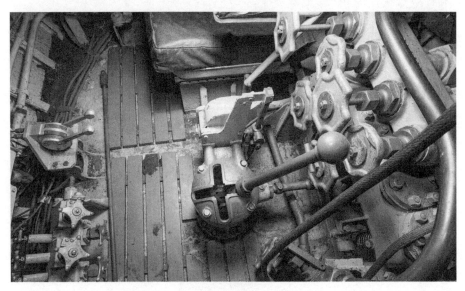

The trim control level. *(Submarine Centre, Helensburgh)*

Left: The planesman's position, showing trim valves. *(Submarine Centre, Helensburgh)*

Below: A periscope-raising wire. The periscope on HMS *Stickleback* has been removed. *(Submarine Centre, Helensburgh)*

Under this compartment was the Q tank, which could be flooded very quickly if the submarine had to dive in an emergency. Also, there was a compensating tank, the aft ballast tank and a small freshwater tank.

WET AND DRY COMPARTMENT

The compartment was entered through a hatch from the control room; a circular hatch in the forward bulkhead led to the battery compartment. There was also a hatch to the casing, which acted as the submarine's main assessment. This hatch also enabled the diver to exit and re-enter the submarine when it was dived.

The water required to flood the compartment so a diver could exit the submarine was carried in a tank under the compartment. The compartment was flooded from this tank and then equalised to the sea pressure so the diver could open the hatch. This arrangement ensured that the submarine's trim would not be badly affected. When the divers returned, the water was drained back into the tank. This compartment also contained the submarine's toilet.

On either side of this compartment were the for'd ballast tanks (port and starboard). The release gear for the side cargoes was also in this compartment.

BATTERY COMPARTMENT

This compartment could be unbolted from the control room/wet and dry section if required, making it easier to change the battery or replace individual cells. The submarine battery had 112 cells.

A fixed periscope enabled the submarine to navigate through the holes in anti-submarine nets that the diver had cut.

There was a buoyancy tank, for'd trim tank, for'd oil tank and compressed air bottles.

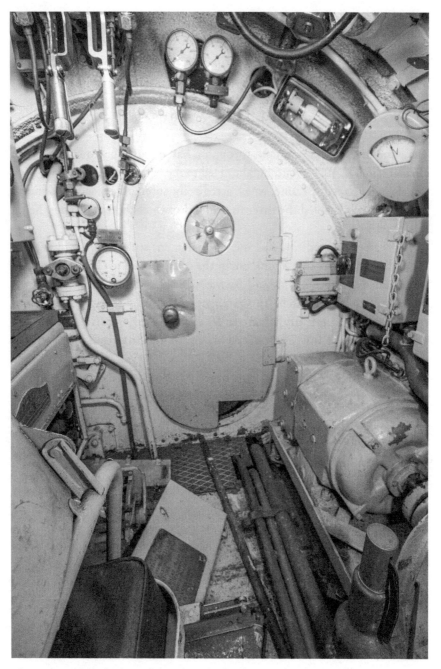

The engine room, looking forward to the hatch into the control room. (*Submarine Centre. Helensburgh*)

LAYOUT

The hydraulic 'plant'. *(Submarine Centre, Helensburgh)*

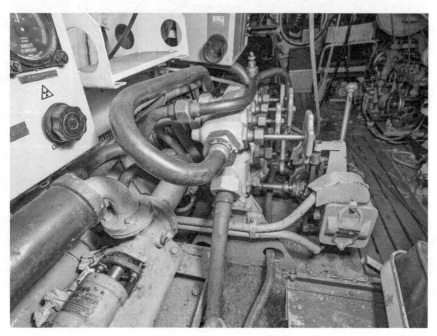

The control room, port side looking forward. *(Submarine Centre, Helensburgh)*

CASING

When running on the surface, the submarine was conned from the casing, which with the very low freeboard, was little more than riding a motorised surfboard for the watchkeeper.

The casing provided the crew with a working platform when coming alongside or leaving harbour. It also gave support and housing for the ventilation trucking, periscope and hydrophone.

Four spring-loaded positioning poles could be raised from the casing to stop the submarine crashing into the target when it released its side cargoes. They also allowed the diver to exit and enter the wet and dry compartment when the submarine was under a target. They were called forwards and backwards antennae.

There was a box at either end of the keel where lead ballast could be placed or removed so the bodily weight of the submarines could be adjusted.

Above: The fuse panel control room. *(Submarine Centre, Helensburgh)*

Opposite above: A fuse panel. *(Submarine Centre, Helensburgh)*

Opposite below: A general view of the planesman's position. *(Submarine Centre, Helensburgh)*

LAYOUT

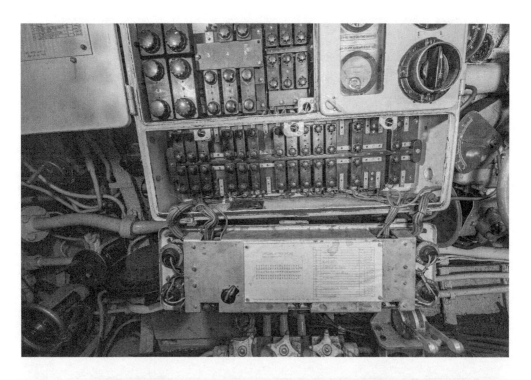

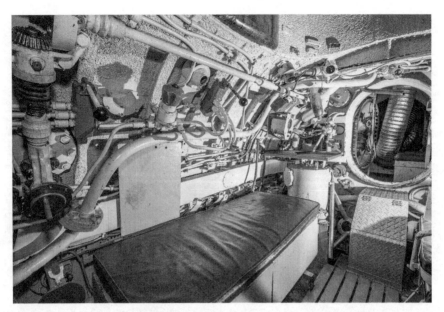

The control room bunk, port side. The hatch to the wet and dry compartment can be seen. *(Submarine Centre, Helensburgh)*

A battery compartment, showing the net periscope. *(Submarine Centre, Helensburgh)*

LAYOUT

The control room, looking forward, port side. The hydrophone can be seen in the foreground. *(Submarine Centre, Helensburgh)*

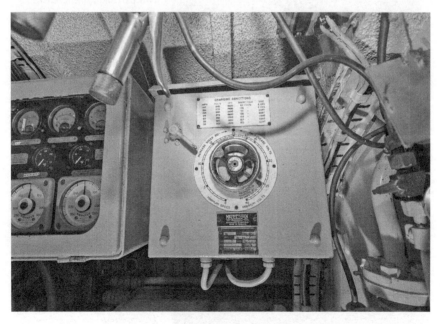

The engine room, showing the electric motor control. *(Submarine Centre, Helensburgh)*

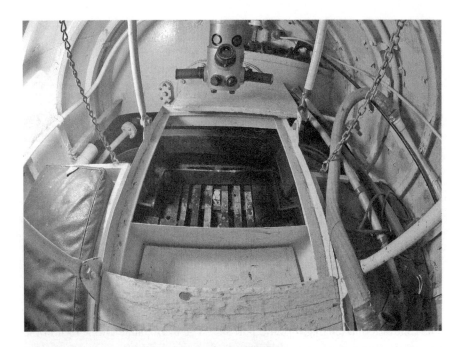

Above: A battery compartment. *(Submarine Centre, Helensburgh)*

Left: The hatch from the engine room into the control room. The ship's gyro compass can be seen on the left. *(Submarine Centre, Helensburgh)*

Opposite above: The gyro compass. *(Submarine Centre, Helensburgh)*

Opposite below: Looking aft from the battery compartment, through the wet and dry compartment into the control room. *(Submarine Centre, Helensburgh)*

Above left: Looking into the wet and dry compartment from the control room, the main access hatch can be seen. *(Submarine Centre, Helensburgh)*

Above right: The wet and dry compartment. *(Submarine Centre, Helensburgh)*

The planesman's position. *(Submarine Centre, Helensburgh)*

LAYOUT

The engine room. *(Submarine Centre, Helensburgh)*

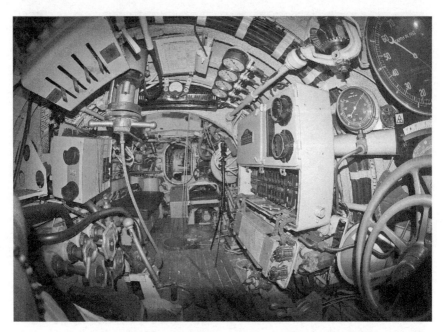

The control room, looking forward. *(Submarine Centre, Helensburgh)*

The engine room, looking aft. *(Submarine Centre, Helensburgh)*

The helmsman's position. *(Submarine Centre, Helensburgh)*

The control room, starboard side. *(Submarine Centre, Helensburgh)*

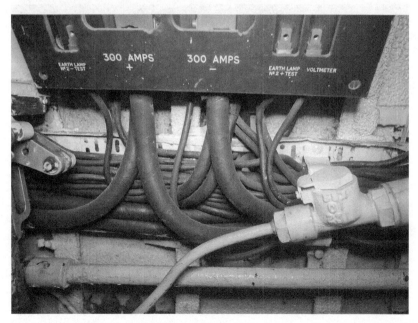

Electrical wiring. *(Submarine Centre, Helensburgh)*

The control room, looking aft, with the helmsman's seat in the foreground on the left. This shows the incredibly cramped conditions. *(Submarine Centre, Helensburgh)*

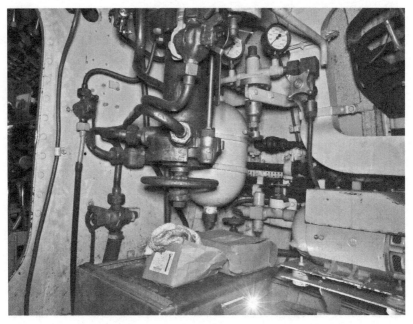

The control room aft, port side. *(Submarine Centre, Helensburgh)*

LAYOUT

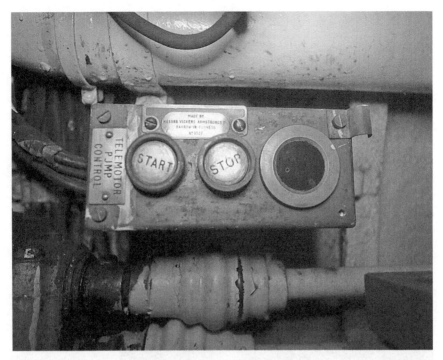

Hydraulic system controls. *(Submarine Centre, Helensburgh)*

Says it all … instructions on the controller panel door. *(Submarine Centre, Helensburgh)*

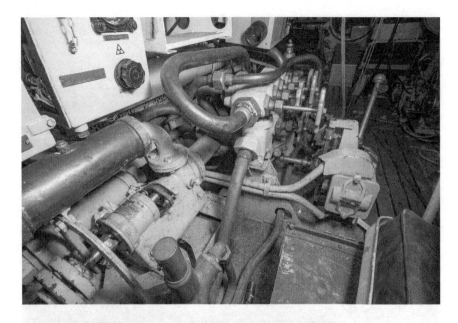

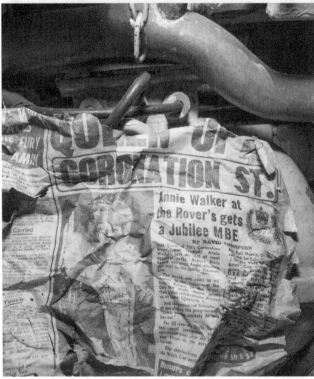

Above: First fitter in gets his pipe in straight. The control room behind the planesman's seat. *(Submarine Centre, Helensburgh)*

Left: A paper that was found in a locker on HMS *Stickleback* when it arrived at the Submarine Centre. *(Submarine Centre, Helensburgh)*

10

HOW THE X CRAFT WORKS

Fundamentally the Stickleback-class submarines worked similarly to their larger sisters; it was just that everything was smaller.

A vessel floats because the weight of the water it displaces is equal to its weight; this causes an upward force known as buoyancy. Most vessels can adjust their buoyancy; submarines are unique in that they can alter their buoyancy to such a degree that they can actually sink. On the surface, the main ballast tanks are filled with air, making the submarine lighter than the water. However, when the submarine dives, the air in the tanks is allowed to escape and the ballast tanks are flooded with seawater. This increases the submarine's weight, overcoming its buoyancy and it begins to sink. There are also several other auxiliary, trim or compensating tanks, which can be filled with water or emptied as required. This helps compensate for changes in the bodily weight of the submarine; for example, using fuel or fresh water would cause the submarine to become lighter. This requires keeping the balance of air and water in these tanks so that their overall density equals the surrounding water.

To help keep the X crafts level when dived, water could be pumped between the bow and stern trim tanks. Water could also be pumped in and out of the submarine using this system. Although not many stores were carried, a small freshwater tank and a few tins of food, the submarine was so small that it was said that even the crew moving about could alter the trim. The submarine could control its angle in the water by the hydroplanes, short movable sets of 'wings' at the stern. A rudder was used to steer, with the hydraulic system operating both the hydroplanes and rudder.

So as not to affect the trim or body weight of the submarine, the water needed to flood the wet and dry compartment was carried in the submarine in a tank underneath it. When a diver was going to exit the submarine, water was pumped from this tank; when the compartment was completely full, the diver equalised the pressure and opened the hatch. When he returned, the water in the compartment was drained back into the tank.

The compressed air needed to blow the ballast tanks was stored in cylinders in various parts of the submarine, so it was ready for instant use; the air bottles were topped up by the compressor when on the surface.

Unlike its larger sister submarines, the midget submarine did not have torpedo tubes; its main armament was two side charges, which could be dropped under a target. These detachable charges contained 2 tons of amatol, an explosive made of 51 per cent ammonium nitrate, 40 per cent TNT and 9 per cent RDX. The submarines also carried limpet mines, which the diver could attach to the hull of its target.

The hydrophone, which allowed the crew to hear noise external to the submarine, was trainable, which allowed the crew to determine the bearing of the sound.

There were 112 battery cells with a capacity of 405AH at a five-hour discharge rate. When the batteries were connected in series, the DC voltage was 220V; in parallel, it was 110V. These batteries reached halfway up the compartment and were covered with a wooden platform that allowed battery maintenance to be carried out and it also provided a space for two bunks.

11

LIFE ON BOARD THE SUBMARINE

These submarines were designed for one task; it was never intended that they would loiter at sea, waiting for a suitable target. Consequently, the accommodation they provided for their crews was at best sparse and they were not intended to house their crews for extended periods.

The X craft operational authorities recognised this fact. The submarines were provided with two crews, one to man the submarine during the transit to the target area. They were then relieved by the operational crew, who carried out the actual attack. X craft operations were measured in days; there was just not the room on board to make longer patrols viable. There was no room to store food, spare parts or even more weapons. Crew resilience was the overriding consideration: how long could they be expected to function efficiently in such cramped and challenging conditions? To help ease these problems, only short submariners were picked to serve on X craft.

In addition to this, submarines are never good sea boats and the smaller the boat, the worse the problem; sea sickness was a common problem. When HMS *Stickleback* was being operated by Swedish crews (see Chapter 12), each crew member had their own bucket and there is no reason to believe that British crews fared any better.

The five crew members had no choice but to live in the control room. Here they kept their watches and relaxed and ate. The control room was their workplace, galley and mess.

There was nowhere to go, the crew really couldn't get up and walk around. There was no room in the engine room and although it was possible to snatch a few hours' sleep in the battery compartment or use the toilet in the wet and dry compartment, there was nowhere else to go.

When the submarine was on the surface, it was usually conned from the casing. With the vessel's low freeboard, this must have been somewhat akin to surfing using a powered surfboard. The hatch of the wet and dry compartment had to be opened while the watchkeeper climbed out and clamped himself to the snorkel (the air intake for the diesel) to provide support and, in rough weather, stop him from being swept overboard. Once he had clipped his safety harness to the ventilating trunk, he could shut the hatch and keep his watch standing on the casing, clinging to a bar protruding from the snorkel close to the voice pipe to the control room.

The commanding officer would be on the periscope and navigating the submarine when dived. The first lieutenant would be controlling the hydroplanes keeping depth, and the ERA would be manning the helm, steering the submarine – that was if he wasn't maintaining the engine. The other two crew members would do the catering and they would relieve the first lieutenant or the ERA when resting.

Each of these watchkeeping positions took great concentration, monitoring the various dials and gauges and moving the rudder and planes to keep the ordered course on depth.

A portable table could be set up in the control room. Sometimes the crew played chess on it or the submariners' unique version of ludo, 'uckers'. However, such was the passion generated by the game that some captains did not allow their crew to play it during the war as they felt it was too exciting and the crew might use up too much oxygen in the heat of the moment.

The crew mostly ate tinned food, which they heated on a small electric hot plate. Their basic diet was meat and two veg, beef stew, baked beans and the Navy's chocolate drink, known as 'Kai'. This was a dark, unsweetened, very hard chocolate that came in a large block wrapped in an oilskin pouch, much as it was in the Second World War. It was grated into a mug and hot water was poured over it. It was often mixed with condensed milk, which also came in a tin. Some crew members would eat it 'raw' and gnaw it directly off the block.

With condensation being such a significant problem, the labels would often come off the cans of food due to the damp atmosphere. As a result, the food had to be identified by painting the name of the contents directly on the can. Apart from confusing the chef, the labels could get into the suctions of trim and bilge pumps.

Unlike larger submarines, the X craft had no form of air purification. If the atmosphere required refreshing, the only choice was for the submarine to surface. When the submarine was dived, the main atmospheric problems were carbon dioxide poisoning and lack of oxygen. The former can cause a crew to suffer headaches, dizziness, nausea, vomiting, rapid breathing and increased heart rate. This can lead to confusion and convulsions; in the worst case, the crew could lose consciousness.

The latter was the greater danger because the sufferer was not aware that anything was wrong; lack of oxygen can cause a sense of euphoria, which is particularly dangerous. In addition, the affected crew may also suffer from changes in skin colour, confusion, shortness of breath, headaches, restlessness, dizziness, rapid breathing, chest pain, high blood pressure, lack of co-ordination and visual disorders. This could cause judgement problems, which could cause crew members to make serious mistakes in operating their craft or plot positions wrongly on the chart, misread bearings and misidentify objects seen through the periscope.

12

STICKLEBACK CLASS OPERATIONAL LIFE

The four Stickleback-class submarines were developed primarily to test harbour defences against the midget submarines the Russians were thought to be using. However, they had comparatively short operational lives regardless of their intended purpose.

On 2 March 1959 in the House of Commons, the MP Captain Henry Kerby asked the Parliamentary Secretary to the Admiralty whether the midget submarines HMS *Sprat*, *Minnow* and *Shrimp* were set for disposal and how long each vessel was in commission. C. Ian Orr-Ewing replied that the three X craft had been built for development purposes. They were now available for disposal and the government was ready to consider offers for them. They were in commission for twenty-one, twenty-three and eighteen months respectively, he added. Captain Kerby also asked the Parliamentary Secretary if he would give an approximate figure for the cost of a vessel of the HMS *Sprat* type of midget submarine. Orr-Ewing answered that the approximate figure, as could be seen in the 1956–57 Navy Estimates, was £93,500.

When the craft was being built, the Admiralty's focus was already turning in a new direction. The US commissioned the nuclear-powered USS *Nautilus* on 30 September 1954 and the Admiralty, particularly FOSM, was very interested in the potential of this innovative form of air-independent power for submarines. The Navy was also accessing the potential of the HPT submarines HMS *Explorer* and *Excalibur*. Again, one of the main drivers was to train surface ships against the high-speed submarines the Russians were believed to be developing.

During their operational life the submarines were based at:

Jan 1955   *Stickleback* at Portland 2nd Flotilla, where HMS *Maidstone* was the depot ship.

Jan 1956   *Stickleback* and *Minnow* at Portland as part of the 2nd Flotilla.

Jan 1958   *Shrimp* and *Sprat* in *Dolphin* with the 5th Flotilla.

While the submarines were based at Portland, they were used to train surface ships to deal with the perceived Russian threat and to test harbour defence tactics.

The submarines made several 'goodwill' visits to towns around the country. I have heard stories about the midget submarines being transported around the country on a special train, which provided a wardroom, senior rates' mess, junior rates' mess and a galley in converted carriages. This, reportedly, caused problems for the crew returning from the obligatory run ashore and then having to navigate the various, unfamiliar hazards associated with berthing in a railway marshalling yard. Perhaps it was during one of these visits that on 30 October 1954, HMS *Stickleback* achieved a new altitude record for a submarine; while travelling over Shap summit in Cumbria on a railway truck, it reached 279m.

The train accommodation might have supported the submarines when they made more conventional visits to British towns by actually sailing to them. The submarines did not provide a particularly comfortable living accommodation, for even a short length of time. And during the visit, the mandatory cocktail party for local dignitaries might not have had the required ambience in the very cramped X craft control room.

The four Stickleback-class submarines were adopted by the Peterborough firm of F. Perkins Ltd, which made the diesel engines that powered the craft on the surface. The crews visited the factory on 1 December 1956 and were presented with a plaque to mark the occasion. The men also received Christmas and birthday gifts and were granted honorary membership of the company's sports association.

In the early morning of 19 July 1956, HMS *Sprat* and *Minnow* sailed from Dartmouth Creek, accompanied by the depot ship *Maggie*, to the entrance of the Exeter Ship Canal. After successfully negotiating the locks, they reached the central basin in the late afternoon. Here they were met by the mayor, in his full ceremonial regalia, and other local dignitaries. The crew were treated to the local delicacy, oggies (Cornish pasties), which were washed down with copious amounts of rough cider. During the short visit, a cricket match was arranged, for which the local team were resplendent in their pristine whites, while the submariners wore an assortment of number eights (working clothes for naval personnel, consisting of a shirt and trousers) and boiler suits. Apart from their unconventional kit, the crews were further disadvantaged by the cider, which was readily available throughout the match.

HMS *Sprat* visited Goole, then sailed to Leeds, Wakefield and York using the Aire and Calder and the Calder and Hebble Canals. It also visited Hull during this trip.

HMS *Stickleback* was transferred to the Royal Swedish Navy. A ceremony was held at Portland Dockyard on 15 July 1958, where she was accepted by Commodore O. Kroltstedt, the naval attaché at the Swedish Embassy in London. The operational and passage crews of the *Stickleback*, both Royal Navy and the Royal Swedish Navy, commanded by Lieutenant D.J.D. Strange. RN and Lieutenant P. Malmgren were inspected by Commodore Kroltstedt, and the Royal Navy provided the guard. Earlier in the year, the Swedish submarine crew had visited Portland for training on the craft.

Once the submarine arrived in Sweden, it was renamed *Spiggen* (Swedish for 'stickleback'). There it was used to test harbour defences. The combat diving teams deployed from *Spiggen* were specially selected, trained and equipped for physically, psychologically and politically demanding situations. Their main task was to scout and carry out sabotage in all conceivable environments and climates. The combat teams believed they were so successful in attacking the large surface ships that the embarrassed senior naval officers had little choice but to sell the midget submarine back to the UK.

Lieutenant Commander Ingemar Curtsell, Swedish Navy (Retired), first went to sea on *Spiggen* as a young trainee officer. He was to return to the submarine several years later in command. Below is his account of his time with *Spiggen*, starting with his first trip on the submarine:

> During this trip, the submariners were able to make a trip submerged with '*Spiggen*'. During this time, Pelle Wide was the Captain. He was also a Navy Seal. Obviously, Pelle wanted to make a proper dive and as it was winter, the sea was covered with ice, although not very thick. We therefore went below the ice and Pelle demonstrated how the sub could make herself light and

raise the hydraulic 'legs' she had on each side to hold the sub below ice in a fixed position by making her light. On each side, *Spiggen* had the possibility to carry two detachable 2-ton side charges, which then could be released. The same principles would be used under a ship's hull. After completing this exercise with *Spiggen*, we all realised that we had selected the right type of naval branch.

This brief recap down memory lane in 'Swinglish' has been established to summarise some of my own experiences from the Navy Seal and HMS *Spiggen* period.

Spiggen was purchased by the Swedish Navy at the same period as the Navy established its own Navy Seal programme. This then became a joint operational unit managed by the diving school's commanding officer.

The Diving school trained both Navy Seals and Hard Hat divers, of which many also took part in the Navy's Deep-Sea Diver programme down to 150m from the Navy's Submarine Rescue vessel Belos 2 and initially from the Rescue submarine URF down to 250m (300m capability).

In Sweden, Navy Seal divers began training at the Hårsfjärden Navy Diving School in 1955. The Navy's Seal operations were discontinued in 1979 and since then transferred to the Amphibious Group and sometimes part of the integrated Special Operation Group (SOG).

Swedish Naval officers Jan T. Sundlöf and Rolf Hamilton received training among others later with the US Navy SEAL programme.

The Swedish Navy Seals were part of the Submarine Flotilla and therefore most of the trainees were selected from regular submariners in addition to enlisted volunteers doing their military service.

The winter training was conducted together with the Mountaineering Ranger School in the Northern forests of the country. Some of the Navy Seals also participated in the annual 80km Wasa cross-country ski race and some went through the Army Para training.

Training in close combat was not just ordinary infantry training. It was conducted by Allan Mann, a Swedish WW II veteran of the British Special Operations Executive, where he also participated in the Dieppe raid with gliders. He showed us British instructional films in close combat and subsequently instructed us in various methods. We certainly learned a lot from his British experiences.

Navy Seals had various transportation means. This could be deployed in pairs from Heli, Beach & Recovery from speed boats, canoeing in kayaks or by larger submarines. We started lock-outs from torpedo tubes followed by the use of newer submarines' rescue chambers and finally from dedicated Navy Seal lock-out chambers.

Then we got a new 'tool'.

Spiggen was a direct development of the RN wartime X craft, which were designed to penetrate enemy harbours and ports and attack shipping. Although other roles were found for these versatile little submarines. *X-51* was ordered on 6 September 1951, launched in July 1954, and completed on 5 June 1955.

She served with the Royal Navy until 1957, having been named HMS *Stickleback* in 1955.

Her armament was two detachable side charges, each of two tons, which could cripple the largest ship. *X-51* was used in the development of the 'Cudgel' nuclear mine, which could have been laid in 'enemy' territories. The programme was abandoned in 1956.

She was sold to the Royal Swedish Navy in 1958 after a refit and renamed *Spiggen* (*Stickleback*) and used for training purposes. She was returned to RN 1976.

Here she moored alongside her 'mother ship' *Kanholmsfjärd* at a temporary location in the Swedish archipelago during summertime.

However, it was not always nice and sunny in this little 'Arctic' country, so we continued to dive in icy conditions, not least to get 'snorkelling' money from the Navy's 'bean counters'.

All the Spiggen Commanding Officers were Navy Seal trained as well as fully trained submariners with later positions as Commanding Officers on board normal-size submarines. Some of the COs were also Hard Hat and Deep Bell diver trained.

Years	Name	Remarks	Years	Name	Remarks
1958–59	Per Malmgren	My Bell diver partner	1963–64	Jan Sundlöf	US Navy UDT [Underwater Demolition Team], Bell diver
1960	Palle Sandberg		1964–66	Peter Wide	Hard Hat & Bell diver, who pushed me into this adventure

HMS *Sprat* alongside in Wakefield. (*Author's collection, via Wakefield Sea Cadets*)

HMS *Minnow*. (*National Museum of the Royal Navy*)

1960–61	Gunnar Rasmusson	My relative	1967–68	Björn 'Lappen' Mohlin	Hard Hat & Bell diver
1961–62	Åke Leion		1969	Ingemar Curtsell	Para, Hard Hat & Bell diver
1962–63	Ulf Drake af Hagelsrum	My 'boss' on 'Draken'	1970–76	Muskö Shipyard	Refit & ready for delivery

The Chief Engineer on board *Spiggen* was Chief Petty Officer 'Alex' Alexandersson since the acceptance period with RN in UK up to the final decommissioning at Muskö Shipyard in Sweden. He was a true legend and took care of the *Spiggen* machinery like a small baby. *Spiggen* was always 'dive able' thanks to him.

He may also be a legend in the RN Chief Petty Officers' mess room during his time over in the UK preparing *Spiggen* for shipment.

The Brits being champions in establishing all kinds of games had one developed where you drunk a pint of beer with a glass of whiskey floating in one go. Times were logged and the quickest won.

Alex was a small guy, so the Brits were laughing when he accepted the bet against the big British sailors. He had obviously been training in secret.

Surprise, surprise he beat them all and set a RN world record, the story tells.

It was a good initial team which started this new 'tool' now called '*Spiggen*'. The first *Spiggen* Commanding Officer Per Malmgren was later my Deep Bell Dive partner when we started the bell diving programme in 1972 on board the Submarine Rescue vessel Belos 2.

Alex sometimes went very quiet in the stern engine room of *Spiggen*, which times he probably was training for a new world record. What did we know? The Chief was the Chief.

All I can say is that the Swedes tried very hard to maintain the same high standards as the Brits in all mini-sub aspects.

I went through the '*Spiggen*' training under 'Lappen' Mohlin as the Commanding Officer. 'Lappen' was quite a relaxed and low-key person and we got along well. In normal operations, you needed at least three operators for shorted durations.

- 1 Navigator, which normally was the Commanding Officer.
- 1 Chief Engineer, taking care of all the machineries.

- 1 Heading and Depth Controller, who also operated the 'snorkel'.
- 4 Passengers in the forward battery room normally Navy Seals.

For longer operations, you might need to increase the team, so you could go two shifts in particular for navigation and steering control. Alternatively, you dropped to a bottom position and took some naps.

As a future Commanding Officer, I had to know the submarine's main machinery in detail, so back to 'smyga' (tracing) pipes and cables. The engine was quite simple, and we had a facility to 'snorkel' (take in air via a mast) to run the diesel engine submerged.

In this mode, you had to be careful so no water went through, although you had a flap, which was supposed to close if the mast went under. This allowed you to charge the batteries in submerged mode.

The steering system was equally simple and could be controlled by one operator. Otherwise there was a proper periscope and very basic navigational aids. Trim and ballast were controlled from a typical British designed old-fashioned unit. It worked and was simple.

The auxiliary system such as the toilet needed a separate course and the same went for the system of transferring divers out and in, obviously also through the toilet space.

Outside the hull, we had four arms with a flat support at the end, which could be lifted hydraulically from the inside and be placed under a ship's hull to keep the sub in position for dropping its load of explosives.

The explosives were contained in a package on each side, also hydraulically releasable.

At the front, you had a cutter for the purpose of cutting through submarine protection nets normally installed around a moored surface vessel to hinder sneak attacks from subs. This was also hydraulically operated but most of the time a diver had to lockout to assist.

The design was based on the experiences gained in particular during the Tirpitz attack, being the most famous one in the history of X craft. These stories we obviously read about in detail.

The Brits had given us a valuable tool with a lot of good thinking behind.

So finally, I got my sub command and this time on board '*Spiggen*'. She was not very big but what a 'tool' to play around with for a young submariner.

At this time, we got extra money each day we had 'snorkelled' as well as when we had made an air dive or a Closed-Circuit Oxygen dive as fully trained Seals. Obviously, we 'snorkelled' alongside the quay every day and we made as many dives as possible to augment our meagre basic naval salaries.

In case of winter and ice this did not stop us. We just dove at the quay and sometimes a trip below the ice for a short trip around the bay. Full speed ahead and a sharp angle brought us back up. Obviously, we had studied the British Tirpitz operation and the ice was a good replacement for a ship's hull.

Probably the Financial Controller at the Submarine Flotilla did not have a clue of what we were doing, and strict project controls were not yet a way of life in the Navy. What do I know as long as we got some extra money and could have some fun in the meantime?

We were in business.

Spiggen's home base was located in the Hårsfjärden area at our own little islands and therefore perfectly suited for relevant training purposes days and nights.

We even had our very own army-trained attack dog during Seal sneak exercises to make it as realistic as possible.

The 'Spetnaz' track was built after one we saw at an invited visit to a camp outside Kiev in Ukraine during a less cold period between the entities. Further details can be obtained in dark rooms without recorders.

In case of operations at the Swedish West coast or other places, *Spiggen* was split in two and transported by train or ferry and bolted together at the arrival. That was a convenient logistic arrangement and allowed us to participate in various planned exercises at key areas in the country.

The *Spiggen* crew was always a flexible bunch of Submariners/Seals.

Navigation was controlled by sightings through the periscope during normal transits submerged.

During sneak attacks, you tried to avoid having the periscope up and then the navigation was performed in another somewhat orthodox manner.

The approach route was carefully calculated, normally having identified suitable rocks, or submerged clean walls on the islands within the intended route. You commenced your attack with low speed, calculated the time to the first island, and when you slightly hit this island, took a new sighting through the periscope and then the next and the next.

It was normally appropriate to take a quick look at some stage to verify your dead reckoning calculations, in particular if there was any prevailing current.

After the attack on the target, you would normally just get the hell out of there and check your position at a suitable location well away.

Back at the base it was a beer or two to celebrate another victory over the 'surface boys'.

HMS *Shrimp* sailing past HMS *Dolphin*. (*National Museum of the Royal Navy*)

HMS *Sprat* at Blackfriars Bridge. (*National Museum of the Royal Navy*)

We had two pairs of Seals in the forward battery room. One at the time locked out through the combined entry chamber, toilet, and lock-out space. One finger shown at the small window to notify inside that you were ready and when out one finger at another window confirming you were ready and waiting for number two to exit. When both out and connected start swimming under water silently to your target and navigating with your handheld compass and keeping around 8–10m water depth.

After having completed your 'sneak attack' return the same way and into the sub battery room. Time to sail home.

This was real submarine work and we all loved it.

We now started to get operational on the submarine escape training and therefore a free ascent campaign took place from '*Spiggen*' in the sheltered bay at 'Vitsgarn'. Normal survival suits were used.

Trained divers from the submarine flotilla commenced doing free ascents at shallow depths from the escape chamber on board '*Spiggen*' followed by ordinary crew members under full control of safety divers.

This went very well, and the experience was in favour to continue this. At this time, the financial geniuses in the Navy seemed to have woken up to the idea of cost cutting. When a bunch of 'surface vessel' deadbeats then jumped into the equation, the first thing to be cut was our old '*Spiggen*', although still diveable.

We received the message with great sorrow, not only the actual personnel at the Diving School but all the old-timers as well. We were subsequently requested to start preparing for her final demobilisation and steam to our nearby shipyard at the 'Muskö' mountain based and nuclear secured shipyard.

Politicians and Naval bureaucrats never liked operational 'mavericks' and so it has always been in all Navies.

Obviously, you had to show the Navy flag for the civilians now and then and they normally loved our tricks and therefore showed up in large crowds.

'*Spiggen*' was at that time an exotic animal and its crew was considered as mavericks or maybe among some as bloody idiots.

One trick we used to do was to place one man in full uniform in the lock-out chamber breathing through a 'bibs' supply and when the sub surfaced, he would stand up and make his salute to the crowd's applause.

What did you not do for the 'civilians', or maybe we liked it as well. What do I know?

Accurate navigation at this time was obviously of interest to the Swedish Navy and *Spiggen* was a cheap test vehicle for such a trial.

Navy Seal HQ to Stockholm inlet at a famous lighthouse called 'Blockhusudden', where we always had a Schnapps and a brief submarine song before entering the port.

Two clever guys from ASEA (later ABB) entered the story and assembled required equipment in the forward battery room.

Mission briefing conducted and we started to steam on the surface.

Normal cabin service was executed and coffee and cakes delivered to the Battery room.

Finally we arrived after several hours and the 'pointy-heads' declared that we were only 250m outside the actual position.

We therefore considered this exercise to be a 'great effort for mankind' comparable to landing on the moon.

Some months later we were approached again by ASEA and now from another department. This time it was to evaluate a 'Fuel cell' concept under development in their Laboratory.

Obviously, this was also of great interest to the Swedish Navy submarines, so instructions arrived to co-operate. Two new 'pointy-heads' descended on our little *Spiggen* and declared the space adequate after a brief inspection.

We were all set to go for this new adventure without really thinking too much about potential complications but trusted the 'brains' involved to take care of such.

However, one day we got some information through the grapevine that the Laboratory had an explosion and the project was delayed for the time being.

That was the time I started to look more carefully at new scientific developments, especially if including myself.

This was probably the closest we came not being able to re-deliver *Spiggen* to the Brits in a suitable shape.

Everything on board was properly checked and Alex the Chief Engineer made ready for her final trip, including a farewell dive on the way down to the yard.

The maximum design depth rating was 90m, but this was reduced gradually for safety reasons to 60m. However, all three of us decided to make a final 90m dive and split a bottle of Champagne at the bottom.

HMS *Sprat*, with two crew members on the casing. (*National Museum of the Royal Navy*)

The crew of HMS *Stickleback*. (*National Museum of the Royal Navy*)

HMS *Stickleback*, under way in the Solent, 1955. (*Frank Dutton*)

So, decided, we went down carefully and already at around 70m we could hear the hull cracking a little, and when finally, at 90m small leaks showed up. SHIT. We quickly drank the Champagne and surfaced. That was the end of it, and we had survived another one.

At the yard, we presented the sub in correct uniforms. An era was sadly at the end.

The yard did a final refit, and she was packed in for travel back to her original home in UK.

After being transferred back to the Royal Navy, *Stickleback* was put on display at the Imperial War Museum at Duxford. After that, it was placed into storage at Portsmouth Naval Dockyard. In 2018, it was moved to the Scottish Submarine Centre in Helensburgh.

Not to be outdone, HMS *Sprat* also managed to squeeze in a 'foreign' towards the end of its operational life. In 1958, it was loaned to the US Navy and spent almost three months testing harbour defences at Little Creek, Norfolk, Virginia.

USS *Alcor* arrived at Portsmouth on 19 June 1958 to transport the submarine and its operational and passage crews to America. The crews consisted of three officers and six ratings. They arrived at Little Creek on 2 July 1958, where they prepared *Sprat* for operations. The submarine was operated by Royal Navy personnel all the time it was in the US. At the end of August, it returned to the UK, arriving on 15 September 1958.

A US Navy picket boat usually acted as a safety boat while the submarine was conducting exercises. One day these two vessels were sailing towards each other so orders could be passed when, suddenly, the picket boat swung across HMS *Sprat*'s bow. Despite the submarine taking avoiding action, the vessels collided and as a result the picket boat sank; fortunately, there were no casualties. A few weeks later, at the party to mark the end of the submarine's deployment, it was presented with the Order of the Golden Ram, commemorating the first sinking of a US warship by the RN since the war of 1812.

Interestingly, it was the second midget submarine to be loaned to America; HMS *XE9* was loaned to the US Navy in October 1952.

In November 1956, C-in-C Home Fleet, Admiral Sir John Eccles, visited HMS *Minnow* and took a short trip to sea in it.

HMS *Stickleback* surfacing during the Navy days in 1955. (*National Museum of the Royal Navy*)

HMS *Stickleback*. (*National Museum of the Royal Navy*)

STICKLEBACK CLASS OPERATIONAL LIFE

HMS *Sprat* during her Yorkshire visit. (*Stephen Ward*)

Above: X51 (Spiggen) in Gothenburg. (*Lieutenant Commander Ingemar Curtsell, Swedish Navy (Retired)*)

Left: X51 (Spiggen) alongside in Sweden. (*Lieutenant Commander Ingemar Curtsell, Swedish Navy (Retired)*)

PART FOUR

THE SUBMARINE CENTRE, HELENSBURGH

13

THE SUBMARINE CENTRE, HELENSBURGH

During the 1840s, several dissenters held two weekly prayer meetings at the Bath Hotel, Helensburgh. The hotel was owned by Mrs Margaret Bell, the widow of Henry Bell, who was a member of this group. As the membership grew, they moved to the old theatre in the Municipal Buildings.

On 23 November 1843, the group petitioned to become a church, which was approved on 27 March 1844. Shortly after this, the elders decided to build their own church and on 11 March 1845, the foundation stone was laid. Three months later, the Relief and Secession Church opened.

With the union of the Relief and Secession churches in 1847, the church was renamed the Helensburgh United Presbyterian Church. However, by 1860 the congregation had outgrown the church, so once again it was decided to build a new one. This was built on Sinclair Street and opened in 1861.

In 1900 the Free Churches and the United Presbyterian Churches united to become the United Free Church, which required a new name. It was proposed that the church should be called the Central Church but the minister, the Rev. Adam Walsh, argued that it sounded too like a station. His preferred choice, St Columba Church, was eventually chosen.

This building was to become the Tower Digital Arts Centre and the Submarine Centre, Helensburgh.

In 2012, local resident and entrepreneur Brian Keating suggested building a new visitor attraction in the centre of Helensburgh that would supply the town with a new and much needed visitor destination, and undoubtedly help the local economy. He had previously developed the golf resort Machrihanish Dunes on the Kintyre Peninsula before going on to establish a destination marketing organisation for the Helensburgh and Lomond area, which became Visit Helensburgh. He proposed two potential projects for the proposed visitor attraction. The first proposal was a Television Museum to celebrate the Helensburgh-born John Logie Baird, the Scottish inventor and electrical engineer who produced the world's first working television system in 1926. The second proposed project was for a Submarine Museum that would honour the proud history of the Royal Navy's Submarine Service and would also recognise the contribution made by the naval community to the local area from their nearby base at Faslane. This became the chosen option.

With the help of Gareth Hoskins, one of Scotland's leading architects, Keating developed a business plan and outline drawings for what would become the first submarine museum in Scotland and presented them to the council and the Royal Navy. The plan called for £4 million to be raised to develop a state-of-the-art seafront building, which he expected would draw national attention and visitors to Helensburgh as well as bring the naval and the civilian communities closer together. At the same time, assisted by the Royal Navy, Keating held meetings with several companies that were involved in the nuclear submarine program to raise financial support for the museum, while persistently urging the local council for assistance. He formed a committee to help coordinate the project, with representatives from the council, Royal Navy, Chamber of Commerce and the Submarine Association.

Unfortunately, the refurbishment of the town centre and seafront was delayed and so the council were unable to give any reassurances as to time scales or even whether or not the site designated for the new museum would be available. At this stage, there was a very real possibility that, because of the uncertainty regarding the availability of the building site, the submarine that was intended to be the main display wouldn't be available. This was due to the tight time frame of when it would be available for transport to Scotland. As all the funding was linked to this building and realising that losing the submarine could cause the collapse of the whole project, Keating had little option but to try to find a new building to house both museum and submarine. To save the project, he personally acquired the St Columba buildings in 2015, intending to use the church hall behind the church to display the submarine.

However, once installed in her new home, it emerged that the submarine, *X51*, would be too small and cramped to allow visitors the opportunity to explore it. Keating wanted to develop a true and meaningful learning experience that would be worthy of the museum and provide the visitors with an interesting and memorable experience that would not only explain how the submarine worked but also detail its history and operational life. He decided that the best way to show the inside of the submarine was to project it onto the outside (he had some experience in technology, having previously worked for Apple). To achieve this, he recruited a talented film editor and a 3D animator who spent the next eighteen months filming each major object and instrument in the submarine's interior. Some 3,500 objects were captured, digitised and then compiled in such a way that the inside could actually be projected on to the outside in high-resolution video, which could then be animated to show how the submarine worked. This unique presentation was submitted to the 2019 Museums and Heritage Awards' 'Innovation of the Year', where it was highly commended.

Commander Bob Seaward RN (Retired) was invited to be a steering committee member when the Submarine Centre concept was first raised. These are Bob's recollections from the early days of the project, written just before it opened:

> There are few submariners and their families that will have missed out on the 'Faslane experience'. It is more than one hundred years since the Gareloch became favoured for submarine trials and, by 2020, all our submarines will be based on the Clyde. In the main, the Submarine Service has been accepted by Helensburgh and [a] generally comfortable relationship has existed. Currently, the Base is growing to accommodate all of our nation's submarine assets. With an expansion of recreational facilities within the complex, this has undoubtedly reduced the need for young submariners to 'run ashore'. This reduced footfall means that Helensburgh is less of a 'Navy town' than previously and, with the Base out of sight and out of mind, for many residents there is no direct link. Like towns throughout Scotland and the UK, Helensburgh, has suffered from lack of investment over the past two decades and has lost its appeal to the traditional visitors from Glasgow and beyond, who used to enjoy their excursions 'Doon the Watter'.
>
> In 2012, the town was approved for major enhancement through the CHORD (Campbeltown, Helensburgh, Oban, Rothesay and Dunoon) project, intended to improve the visitor experience with stylish walkways, transformation of the seafront, the pier development and a complete remodel of the tired and limited Colquhoun Square. In a move to reunite Helensburgh with the Submarine Community and to boost the almost non-existent visitor experience, under the direction of a local entrepreneur and with the support both of the Navy at Faslane and Argyll and

Bute Council, a plan was developed to construct a brand-new Scottish Submarine Centre visitor experience alongside the 'soon to be redeveloped' Helensburgh pier. The £4 million building was to be mounted alongside the western edge of Helensburgh pier; effectively a glass box resembling the outline of a submarine conning tower, supported on stainless steel piles, allowing a degree of vertical movement with the tide. This was a very ambitious design, intended to capture the imagination. The building would have been clearly visible from land and sea; its submarine profile enhanced by working periscopes held in store for the project. Inside the building, the focal point was to be the former HMS *STICKLEBACK / X51*, a submarine built in 1952 and operated by the Royal Navy for only a few years, before being sold to the Swedish Navy. After a relatively short time in service, the Swedes decommissioned *X51*, subsequently returning her to the UK, where she became part of a general display at the Imperial War Museum, Duxford.

The Royal Navy Submarine Museum Director discovered that the submarine was to be relocated from Duxford, that there were no established plans for its future and so started the process to move the craft to Helensburgh. Unsurprisingly, the plan for the redevelopment of Helensburgh pier is still to be finalized, so the construction of the highly innovative Scottish Submarine Centre was put on hold. The project naturally had built up a considerable head of steam and, to avoid another exciting plan being nipped in the bud, the loss of X51 and a halt to fund raising, alternatives were sought. Fortunately, the former St Columba's Church on Sinclair Street was up for sale and there had been no takers. A decision was made to keep the momentum going by purchasing this site and turning it into a major, multi-function attraction to entertain both residents and visitors to Helensburgh. In short, the main Church was developed into a state-of-the-art cinema, media studies classrooms and a live performing arts centre. Part of the building was developed into a computer training centre, popular with both young people and senior citizens wishing to enter the world of social media. A second bijoux cinema was opened in early 2016. All of these assets are very well supported and are now financially self-sustaining.

What about the Scottish Submarine Centre [SSC]? The very large hall within the Church footprint was earmarked exclusively for X51 and work began to prepare this building to take the craft in 2014. Down south, the

National Museum of the Royal Navy [NMRN] emerged, encompassing all the maritime visitor attractions under one body. Relations between the NMRN and SSC management were established and the somewhat lengthy process of renegotiating the loan of *X51* to Helensburgh was set in train. The SSC was prepared for the submarine's arrival by the summer of 2015; the agreement was signed and the submarine subsequently arrived in Helensburgh in September 2016. The temporary wall was opened up and *X51* was inserted into the building stern first. The two halves were supported by the incredible Versilift tractors and the process of manoeuvring the sections together was begun. Thanks to the incredibly skilled operators, the mating took place without too much effort and a team of volunteers from HMS *AMBUSH* manually sweated 120 stainless steel bolts to make the job permanent.

The next task was to erect the steel pillars upon which the submarine would sit some 3 metres above ground. The pillars were concreted into their supporting holes and after a period of hardening and curing, the Versilifts raised the submarine onto the pillars. A team of Submariners Association volunteers took control of the de-rusting and painting of the craft and, for the time being, she awaits her new role as a media star. It remains for the internal walls to be prepared, ready for the fitting of projection screens. The electrical conduits will then be laid along with the underfloor heating system and topped off with a super-smooth floor finish. Placement of 28 media projectors and associated servers will finish the installation and a period of testing and tuning will ensure that the experience works. Visitors will be able to interact with submariners past and present and discover what drove these young men to join this exclusive club. They will be reminded of the sacrifice made by submariners, as recorded on the digital Memorial Wall. As time goes on, this record will be expanded to contain more in-depth detail of individuals, obtained both by research and from contributions left by visiting relatives. Local secondary school pupils are involved with this task that will be ongoing for years to come. Finally, the hall will be transformed into an underwater experience, where the workings of a submarine will be demonstrated, before X51 herself leads visitors on a journey into submarine history, to entertain, inform, educate and to stimulate the imagination. The SSC is not a traditional submarine museum, rather a state-of-the-art media centre. We look forward to entrancing our visitors when the doors open in August.

Both the Royal Navy at HM Naval Base Clyde and the Argyll and Bute Council supported this project and it was hoped that this new historical and cultural visitor attraction would be open by July 2014, in time for that summer's Commonwealth Games, which were to be held in Glasgow. In addition, it was hoped that the centre would attract more than 10,000 visitors a year to Helensburgh, which would benefit not only the town but the surrounding area as well.

Unfortunately, the project's opening was delayed due to funding issues and, in 2016, Keating wrote to George Osborne, the Chancellor of the Exchequer, seeking £659,000 to complete it. His request was granted in the 2016 budget. Nearly five years after the initial idea of building a memorial to Scotland and Helensburgh's involvement in submarine history, it was with a deep sense of relief amongst all those involved when, on 11 November 2017, the centre was finally opened by Rear Admiral M. Gregory RN (Retired).

Keating, now chair of the Scottish Submarine Trust, said: 'It will be opening on Friday, with no ceremony. There is a feeling of relief that we're ready to have the centre open, but we're moving straight into the next phase. We've booked a booth at the VisitScotland Expo this spring and will be meeting 1,200 tour operators and talking to them about how we can help encourage more visitors to Helensburgh. We want the whole town to see the benefits of the new centre. It's not just a case of opening the doors and hoping people will come in.'

On 10 July 2019, the centre welcomed a royal visitor, HRH Princess Anne, the Princess Royal, who had attended the fiftieth anniversary celebrations of the Continuous at Sea Deterrent programme at HM Naval Base Clyde earlier in the day. The Princess's visit marked the official opening of the centre. Apart from seeing the presentation, she met local tradespeople and volunteers who had worked hard over the previous few years to complete the centre and bring the *X51* presentation to fruition. Members of the Police Scotland Young Volunteers group provided a guard of honour.

Visitors enter the centre via the newly built extension, which forms a reception area, where they can browse a line of submarine-related products in the small shopping area. In addition, it is planned that a café will be built off the reception area. Most centre guides are ex-submariners with a wealth of experience in the submarine service and its history.

THE SUBMARINE CENTRE, HELENSBURGH

HMS *Stickleback* arrives at her new home in Helensburgh. (*All photographs in this section courtesy of Ron Rietveld*)

The submarine was transported to Helensburgh in two sections, which required a great deal of planning and very careful and precise handling to get them into the exhibition hall and lift them onto their plinth.

THE SUBMARINE CENTRE, HELENSBURGH

X3 TO X54: THE HISTORY OF THE BRITISH MIDGET SUBMARINE

After a short introductory talk, visitors are led into the main exhibition hall, where they are met with the impressive sight of *X51*, raised 3m above the floor.

There is then a forty-minute audio-visual presentation, which covers the history and development of the X-craft midget submarines, the Cold War, the building of the Stickleback class, the period when HMS *Stickleback* was loaned to Sweden and how the submarine works. During the presentation the various internal items are projected onto the hull so, effectively, visitors can see inside the submarine. The presentation then shows how HMS *Stickleback* was brought to Helensburgh and fitted into its new home and how the hall was converted and prepared to accept the submarine.

The exhibition is constantly evolving; it is hoped that, in the future, visitors will be able to 'walk through' the submarine using virtual reality headsets. In addition, other submarine-related presentations are planned, including one on *K13* and another on the history of HMNB Clyde.

Once the submarine was in place the new entrance was built. *(Submarine Centre, Helensburgh)*

As one of Scotland's most unique venues, the centre is used for various other purposes; its innovative use of immersive technologies makes it the ideal setting to host multiple meetings and cultural events. It has been used as a meeting place for the West of Scotland Branch of the Submariner's Association, an annual dinner location for the Submarine Coxswains' Association, and has held several concerts. In November 2017, the hall was used for a Remembrance service, where Scott McGinley read a letter that was sent to a grandparent of one of the centre's volunteers, informing them of the death of their son at Gallipoli, while the Helensburgh String Quartet played an accompanying recital. During the 2022 Remembrance weekend, the presentation was shown several times a day, with a recorded accompaniment.

The hall has also been used for an anniversary celebration, two book launches and various art exhibitions, while Santa's underwater grotto at Coral Cove was a unique and very much-enjoyed event. It has also shown films; perhaps not surprisingly, the Beatles' *Yellow Submarine* was the first to be shown.

THE TOWER ARTS CENTRE

Having acquired such a large building, Brian Keating founded another charity and converted the actual church into the Tower Digital Arts Centre, which is staffed and run by volunteers. It was opened by Holyrood's Cabinet Secretary for Culture, Fiona Hyslop, on 26 August 2015.

Originally it was intended that live performances, including ballet productions, comedy nights, jazz concerts and orchestral performances, would be interspaced by occasional film shows. Unfortunately, the live shows were not particularly well supported, so the decision was made to upgrade the projection equipment to full digital cinema quality and the Tower Digital Arts Centre opened on 15 April 2016 with a screening of Disney's *Jungle Book*.

The centre now has three screens: Screen 1, which seats 150 (and has a bar); Screen 2, which seats 48; and Screen 3, which seats an intimate 24 on sofas.

Centre volunteer Clair Lang single-handedly painted the sub and the 50m of walls with special screen paint. (*Brian Keating, founder of the Submarine Centre*)

HRH Princess Anne talking to Brian Keating. (*Submarine Centre, Helensburgh*)

The centre being used for a photographic exhibition. *(Submarine Centre, Helensburgh)*

Submarine internals projected onto the hull, allowing visitors to 'see inside the craft'. *(Submarine Centre, Helensburgh)*

A recital. *(Submarine Centre, Helensburgh)*

Remembrance Day poppies. *(Submarine Centre, Helensburgh)*

A Remembrance concert, November 2021. *(Submarine Centre, Helensburgh)*

An art exhibition held at the Centre. *(Submarine Centre, Helensburgh)*

Above, right and opposite: Christmas at Coral Cove, 2022. *(Submarine Centre, Helensburgh)*

THE SUBMARINE CENTRE, HELENSBURGH

A piano recital at the centre. *(Submarine Centre, Helensburgh)*

A film being shown in the exhibition hall. *(Submarine Centre, Helensburgh)*

EPILOGUE

The men of the 12th Submarine Flotilla undoubtedly made a significant and crucial contribution to the country's efforts in the Second World War, a contribution that came at a very high cost. Of the thirty-nine personnel from the 12th Submarine Flotilla who gave their lives on active service in their country's defence, only four have marked graves. One is in Tromsø, Norway, for Lieutenant Lionel Barnett Whittam from *X7* and three are in the cemetery at Rothesay; the casualties from the sinking of *XE11* are in Loch Striven. We owe it to these young men, who gave so much yet asked for so little, to remember them and recognise the sacrifices they made. HMS *Stickleback* at the Submarine Centre Helensburgh is a part of the tribute to these craft and the men that crewed them but there are other memorials specifically for X craft.

Fittingly, there is an X craft memorial at Port Bannatyne alongside the loch where the submarines were berthed. A plaque that lists all the X craft crew members who gave their lives in the Second World War has been added to the memorial. A little further along the coast, in front of the boatyard, is the Port Bannatyne Memorial Garden, which is dedicated to the men of the midget submarine flotilla; it was unveiled in 2005. The small garden comprises a foliage-covered model submarine that was made using chicken wire. There are hand-painted slates that bear the pennant numbers of all the X craft that were lost during the war.

The 12th Submarine Flotilla Memorial Cairn was erected at Garbh Eilean, Kylesku, in the Scottish Highlands, to commemorate the fiftieth anniversary of the flotilla. It is in the car park on the north side of the new bridge and was unveiled in 1993. In 2013, Commander J. Lorimer DSO RN (Retired) laid a wreath at the site to mark the flotilla's seventieth anniversary.

The plaque on the cairn reads:

This cairn was erected to
commemorate the 50th anniversary of
the XIIth Submarine Flotilla
10th April 1993

The security of these top-secret operations was
guarded by the local people of this district who
knew so much and talked so little

The silent hills remember the young men of
His Majesty's X craft submarines and human torpedoes
who were trained in these wild and beautiful waters

At the going down of the sea we will remember them:

R. Anderson	Ord Seaman R.N.R.	C. Ludbrook	E.R.A.
G.E. Bonnell	D.S.C. Lt. R.C.N.V.R.	A.D. Malcolm	Sub. Lt. R.N.V.R.
A.J. Brammer	Ldg. Stoker	R. Maplebeck	Able Seaman
P.C.A. Brownrigg	Lt. R.N.V.R.	W.J. Marsden	Lt. R.A.N.V.R.
D. Carey	Lt. R.N.	B.M. McFarlane	Lt. R.A.N.
J.J. Carroll	Able Seaman	R. Moritboys	E.R.A.
H.P. Cook	Lt. R.N.V.R.	I.J. Nelson	Sub. Lt. R.N.V.R.
B. Enzer	Lt. R.N.V.R.	I. Pretty	Able Seaman
R. Evans	Able Seaman	R.W. Pridham	Stoker
G.G. Goss	Sub. Lt. R.N.V.R.	P.D. Purdy	Sub. Lt. R.N.Z.N.V.R.
L.B. Grogan	Sub. Lt. S.A.N.F.,(V.)	M. Rickwood	Ldg. Seaman
K.V.F. Harris	Sub. Lt. R.N.V.R.	I. Sargent	Sub. Lt. R.N.V.R.
A.H. Harte	Ord. Seaman	W. Simpson	Able Seaman
H. Henty-Greer	Lt. R.N.V.R.	A. Staples	Lt. S.A.N.F.,(V.)
E.W. Higgins	Stoker	S.F. Stretton-Smith	Lt. R.N.V.R.
G.H. Hollett	Stoker	T.M. Thomas	Sub. Lt. R.N.V.R.
P.J. Hunt	Sub. Lt. R.N.V.R.	B. Trevethian	Ldg. Seaman
F. Kearon	Sub. Lt. R.N.V.R.	W.M. Whitley	E.R.A.
J. Kerr	2nd Lt. H.L.I.	L.B.C. Whittam	Lt. R.N.V.R.
D.H. Locke	Sub. Lt. R.N.V.R.		

Unveiled by Captain Robert I. Bradshaw Royal Navy
Captain SM 10th Submarine Squadron   31st May 1994

EPILOGUE

On the wall of the Normandy Museum in Portsmouth is a plaque commemorating the two X craft (*X20* and *X23*) that took part in the Normandy Landings (Operation Gambit). They indicated the western and eastern limits of Sword and Juno beaches for the Anglo–Canadian landings.

There are several X craft on display that can be visited by members of the public, and these include two surviving examples of a Second World War X craft.

X24, which attacked the floating dock at Bergen on 15 April 1944 in Operation Guidance, is at the Royal Navy Submarine Museum at Gosport, Hampshire, where it has been on display since 1981. *XE8* (Expunger) is on display at Chatham dockyard. It was sunk as a target in 1952 and salvaged off Portland Bill in 1973.

Not so well preserved are two XE midget submarines that were towed to Aberlady Bay, East Lothian, Scotland, in spring 1946. During the first week of May they were used as targets for aircraft including Mosquitoes and Seafires (the naval version of the Spitfire) in a trial to access the effectiveness of 20mm cannon shells against the hulls of the submarines.

The only surviving example of the Stickleback-class midget submarines is HMS *Stickleback* itself and this is displayed at the Submarine Centre, Helensburgh.

All in their own way are memorials to the men who manned the X craft and they tell the story of these brave men who are no longer here to tell their own story.

May their remembrance be as lasting as the land they honoured.

The war memorial in Port Bannatyne, showing the 12th Flotilla plaque. *(Unless credited, all photos in this section courtesy of Ron Rieveld)*

The plaque to the two X craft involved in the Normandy Landings (Operation Gambit). *(Author's collection)*

EPILOGUE

The X craft garden at Port Bannatyne.

EPILOGUE

The graves of the crew members of *XE11*, which sank in Loch Striven after colliding with a boom defence vessel.

BIBLIOGRAPHY

Akermann, Paul, *Encyclopaedia of British Submarines 1901–1955* (Periscope Publishing Ltd).

Bafgnasco, Erminio, *Submarines of the Second World War* (Cassell and Co.).

Gallagher, Thomas, *The X-Craft Raid* (Harcourt Brace Jovanovich).

Hall, Keith, *HMS Varbel* (Kindle Edition).

Hall, Keith, *Operational Cudgel* (Kindle Edition).

Harding, Leighton, *Tirpitz Nemesis: A Short History of the X Craft Midget Submarines* (Kindle Edition).

Lyman, Robert, *The Real X-Men: The Heroic Story of the Underwater War 1942–1945* (Quercus).

Sainsbury, Captain A.B., & Phillips, Lt Cdr E.L., *The Royal Navy Day by Day* (Centaur Press).

BBC News: 'X men: World War II's midget submarine crews', 22 September 2013.

RN Subs, the website of the Barrow Submariners Association: RNSubs.co.uk

Hansard: Midget Submarines HC Deb 2 March 1959 vol. 601 cc16-7W.

The National Archives: ADM 1/24063 Policy for the Development and Maritime Employment of the X Craft Flotilla.

ACKNOWLEDGEMENTS

I am grateful to the following for their assistance and support in helping me complete this book. Chris Leggett, for his help with the technical aspects of the book. Also, Ron Rietveld for his support and help with the mechanical parts of the book and for allowing me to use his photographs of the Port Bannatyne memorials and the Submarine Centre.

Barrie Downer for allowing me to use Mr Weatherburn's reminiscences of building the wartime X craft and his visit to their Port Bannatyne base, and Tony Prosser for the background information about Port Bannatyne.

Commander Bob Seaward OBE RN (Retired) for permission to use his recollections of the centre's early days.

Special thanks to Brian Keating and Dave Dunbar for their support and advice, and for allowing me access to the Submarine Centre's photographic archive and to make off with the contents.

Kate Braun from Royal Navy Museum Portsmouth for the X craft photos and Charlotte Hawley from the Barrow Dockyard Museum for the photographs of the HMS *Stickleback* build. Stephen Ward and *The Rothwell Record and Advertiser* for the pictures and information regarding HMS *Sprat*'s visit to Yorkshire.

Special thanks to Ingemar Curtsell, Lieutenant Commander (Retired), for his memories of *Stickleback*'s time with the Swedish Navy.

I had the honour and pleasure of meeting and getting to know two X craft men, who served during the Second World War: Adam Bergius and Bill Morrison. Over a number of years I enjoyed their company and listening to their stories, which they allowed me to use. Sadly both gentlemen have now 'crossed the bar'. It's a privilege to be able to share their stories with you.

As always, Amy Rigg and the team at The History Press for making sense of my jumbled and disorderly manuscript, and particularly her patience and allowing me to tell this story.

All MoD quotes from official documents are reproduced under the Open Government Licence v3.0.

I must stress any mistakes in the book are due to my inattentiveness. All the people involved couldn't have been more helpful and showed a remarkable degree of patience towards me and my endless and often inane questions; their contribution to the book has been immeasurable.

ALSO BY THE AUTHOR

Trident: The UK's Submarine Force After Polaris (The History Press, coming 2024)

K13: The Untold Story (Independently published, 2019)

Polaris: The History of the UK's Submarine Force (The History Press, 2018)

HM Naval Base Clyde (The History Press, 2012)

Submarines News: The Peculiar Press of the Underwater Mariner (The History Press, 2010)

HMS Defiance (The History Press, 2008)

Gareloch and Rosneath (Pocket Images) (Nonsuch, 2007)

Submariners: Real Life Stories from the Deep (The History Press, 2006)

Rosneath and Gareloch: Then and Now (Tempus, 2002)

The Clyde Submarine Base: Images of Scotland (Tempus, 2002)

Around the Gareloch and Rosneath Peninsular (The History Press, 2000)

HMS Dolphin: Gosport's Submarine Base (The History Press, 2000)

YOU MAY ALSO ENJOY ...

978 0 7524 5177 0

Polaris is not just the history of the weapons, submarines and politicians: it is the history of those who were there.

978 0 7509 9681 5

David L. Williams examines the origins of maritime camouflage, its natural influence, its effectiveness and its personel.

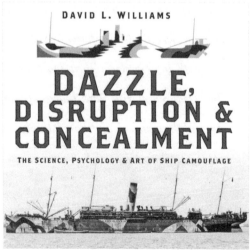

FREE Test Taking Tips DVD Offer

To help us better serve you, we have developed a Test Taking Tips DVD that we would like to give you for FREE. **This DVD covers world-class test taking tips that you can use to be even more successful when you are taking your test.**

All that we ask is that you email us your feedback about your study guide. Please let us know what you thought about it – whether that is good, bad or indifferent.

To get your **FREE Test Taking Tips DVD**, email freedvd@studyguideteam.com with "FREE DVD" in the subject line and the following information in the body of the email:

    a. The title of your study guide.

    b. Your product rating on a scale of 1-5, with 5 being the highest rating.

    c. Your feedback about the study guide. What did you think of it?

    d. Your full name and shipping address to send your free DVD.

If you have any questions or concerns, please don't hesitate to contact us at freedvd@studyguideteam.com.

Thanks again!